Optical Image Formation
and Processing

OPTICAL IMAGE FORMATION AND PROCESSING

M. FRANÇON

Optics Laboratory
Faculty of Sciences
University of Paris
Paris, France

Translated by BERNARD M. JAFFE

Department of Physics
Adelphi University
Garden City, New York

ACADEMIC PRESS   New York   San Francisco   London   1979

A Subsidiary of Harcourt Brace Jovanovich, Publishers

√ 6310 6735

PHYSICS

ACADEMIC PRESS, INC.
111 Fifth Avenue, New York, New York 10003

United Kingdom Edition published by
ACADEMIC PRESS, INC. (LONDON) LTD.
24/28 Oval Road, London NW1   7DX

Library of Congress Cataloging in Publication Data

Françon, Maurice.
   Optical image formation and processing.

   Translation of Optique, formation et traitement des
images.
   1.   Optics, Physical.   2.   Coherence (Optics)
3.   Fourier transformations.   I.   Title.
QC395.2.F713   535'.2   78−19668
ISBN 0−12−264850−1

French edition published as *Optique.
Formation et traitement des images.*
© Masson, Editeur, Paris, 1972.

PRINTED IN THE UNITED STATES OF AMERICA

79 80 81 82   9 8 7 6 5 4 3 2 1

Contents

Chapter 3 **Diffraction**

Chapter 4 **Diffraction Gratings**

Chapter 5 **Partial Coherence**

Chapter 6   Interference Phenomena in Polarized Light

Chapter 7   Formation of Images. Filtering of Spatial Frequencies by an Optical Instrument

Chapter 8   Holography

Appendix A   Review of Some Elementary Concepts regarding the Fourier Transformation

Appendix B   The Fresnel–Kirchhoff Formula and the Phenomena of Diffraction

Brief Bibliography

Preface

For twenty years, information theory and communication theory have greatly influenced optics. In communication systems the information is generally of a temporal nature; on the other hand, it has a spatial nature in optics, where the question is how an image is formed or transmitted by an optical instrument. The similarity of the phenomena of communication theory and optics has led to a formulation which is practically the same for both and which is based on Fourier analysis.

The introduction of the Fourier transformation into optics was the starting point for a substantial development, both in the formal structure and in the fundamentals. Fourier optics is the modern form of physical optics. It is also called coherent optics, and its principal elements are coherence, holography, optical filtering, and the processing and storage of information. Of course, these developments rest on the fundamental bases of interference and diffraction, which themselves gain much clarity when they are presented from the standpoint of the Fourier transformation. For example, the interference fringes produced by two point sources of light (Young's interference fringes) are given by the Fourier transform of two Dirac distributions. A grating may be thought of as a Dirac comb, and its diffraction phenomena as the Fourier transform of this Dirac comb. This is, once again, a Dirac comb, and in this way one immediately obtains the spectra of the grating.

No matter how one envisages the presentation of optics in the teaching of physics at the University, it seems very desirable to present the basic facts and the recent developments in a form which is both precise and concise. We are therefore led, in the present work, to present physical optics in its modern aspect at the same time that we depart, to some extent, from the traditional modes of presentation. But the appearance of new questions ought to be accompanied by a certain modification of the material taught, and we had thought that the electromagnetic theory, deduced from Maxwell's equations, could be taken up after the fundamental theorems in a course in electromagnetic theory. We thus restricted ourselves to introducing the concept of a luminous vibration.

The plan which we have adopted results from the preceding remarks. The first four chapters are devoted to interference and to diffraction. We next set forth a simplified theory of partial coherence which brings us to a deeper understanding of these phenomena. Chapter 6 concerns polarization and, more particularly, interference phenomena in polarized light. One might perhaps think that this study would be more appropriate in a course in mineralogy. But it would seem peculiar to assume that physicists could ignore the phenomena of polarization, which have so very many applications and which lead us right up to nonlinear optics. As compared with older treatises, however, this chapter has nevertheless been reduced. In Chapter 7, we explain the formation of an image in terms of the filtering of spatial frequencies, filtering which is based on the concept of the transfer function. But the formation of an image may also be conceived of in a very different way, in terms of holography, and this is treated in the following chapter. The use of the computer in the generation of holograms is not forgotten. Interferometry, including speckle interferometry, is the subject of Chapter 9, and holography plays an important role there; thanks to it we may cause interference between waves recorded at different times! Chapter 10 is devoted to the optical processing of information. Filtering, the recognition of shapes, the detection of differences between two shapes, the storage of information, and the processing of images by computer are studied. We conclude with a chapter on certain experimental effects obtained with lasers, particularly second harmonic generation, and on optical communications.

In thus presenting optics in terms of a conceptual background, we have not sought to collect questions that are more or less related and that pertain to different fields of study. An experiment long ago proved to us that a simultaneous study of phenomena of the same sort in different branches of physics will be stimulating only if it is based on a sufficiently deep understanding of these phenomena. In our opinion this condition is not fulfilled at the level of the "maîtrise," and this is why we have made use of the traditional presentation in the hope that the reader, on closing this book, will soon carry on the study of this beautiful science of optics.

Notation

We think that it will perhaps be useful to go into detail about the notation and the basic diagrams, outlines, and plans which will be used. In studying the phenomena of diffraction at infinity (Fraunhofer diffraction), the aperture producing the diffraction is located in a plane passing through the axes $C\eta\zeta$ (Fig. N.1). The phenomenon of diffraction is observed in the focal plane of an objective O. Since the aperture is always illuminated by parallel light, the plane Fyz is the plane in which is found the image of the point source. One constantly studies what happens in this plane, and one rarely considers what happens in the plane of the source. This is why we have chosen the notation y, z rather than the traditional notation y', z'. The distribution of amplitudes in the plane of the aperture is given by a function $F(\eta, \zeta)$. The distribution of amplitudes in the plane where the

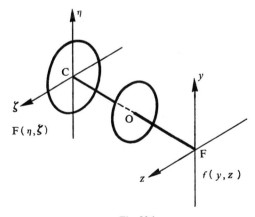

Fig. N.1

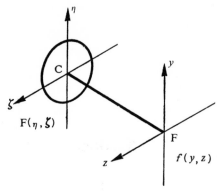

Fig. N.2

diffraction phenomena are observed (the Fourier plane) is represented by a function $f(y, z)$ which is the Fourier transform of $F(\eta, \zeta)$.

We shall employ the same notation and the same schematic diagram in the case of Fresnel diffraction, but the objective O is omitted (Fig. N.2). If one considers an incoherent object and its image as given by an objective, points in the object plane are specified by a system of coordinates having the axes $F_0 y_0 z_0$. As before, the image plane is labeled by a system of axes Fyz (Fig. N.3).

An object plane that is coherently illuminated plays the same role as the aperture that is employed in Fraunhofer diffraction phenomena. It is thus labeled by axes $C\eta\zeta$ and the Fourier plane by axes Fyz (Fig. N.4). Now and then, but only very rarely, one has occasion to

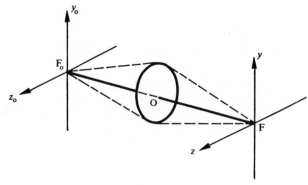

Fig. N.3

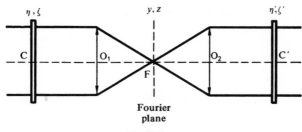

Fig. N.4

consider the plane which is conjugate to the plane $C\eta\zeta$; this conjugate plane is labeled by the axes $C'\eta'\zeta'$.

In holography the plane of the hologram during recording and reconstruction is labeled by the axes $C\eta\zeta$. However, in the case of a Fourier hologram it is the notation Fyz which will be employed. This type of hologram is in fact often employed in the Fourier plane of the assemblies used in optical filtering.

To sum up, we shall use the following notation:

1. Diffraction at Infinity (Fraunhofer)

Plane of the aperture producing the diffraction	axes $C\eta\zeta$
Amplitude distribution in this plane	$F(\eta, \zeta)$
Fourier plane	axes Fyz
Amplitude distribution in the Fourier plane	$f(y, z)$
Distribution of intensity in this plane	$I = f \cdot f^*$

2. Diffraction at a Finite Distance (Fresnel)

Plane of the aperture producing the diffraction	axes $C\eta\zeta$
Amplitude distribution in this plane	$F(\eta, \zeta)$
Plane where the diffraction is observed	axes Fyz
Amplitude distribution in this plane	$f(y, z)$
Distribution of intensity in this plane	$I = f \cdot f^*$

3. Image of an Incoherently Illuminated Plane Object

Plane of the object	axes $F_0 y_0 z_0$
Intensity distribution in the geometric image of the object	$O(y, z)$
Image plane	axes Fyz

Intensity distribution in the true image
   of the object $E(y, z)$

4. Image of a Coherently Illuminated Plane Object

Plane of the object   axes $C\eta\zeta$
Amplitude distribution in this plane $F(\eta, \zeta)$
Fourier plane   axes Fyz
Amplitude distribution in this plane $f(y, z)$
Intensity distribution in this plane $I = f \cdot f^*$

5. Holography and Optical Filtering

Plane of the hologram during recording and
   reconstruction   axes $C\eta\zeta$
Amplitude in the plane of the hologram $F(\eta, \zeta)$
Plane of a Fourier hologram   axes Fyz
Amplitude in the plane of a Fourier hologram $f(y, z)$

In order to simplify the formulas often used in the study of diffraction phenomena, we have set

$$\frac{\sin x}{x} = \text{sinc}(x).$$

Optical Image Formation
and Processing

CHAPTER 1

Luminous Vibrations

1.1   Monochromatic Light Waves. Temporal Coherence

From electromagnetic theory, light is due to the simultaneous propagation of an electric field and a magnetic field. Various experiments show that the vibrations of the electric field may represent the light wave in the regions of space where the light propagates.

Consider an isotropic dielectric substance in which the speed of propagation of the electric field $\mathbf{E}$ is equal to V. Starting from Maxwell's equations, one may derive the equation of propagation:

$$\nabla^2 \mathbf{E} = \frac{1}{V^2} \frac{\partial^2 \mathbf{E}}{\partial t^2} \tag{1.1}$$

where $\nabla^2 \mathbf{E}$ is the Laplacian of $\mathbf{E}$ and t is the time. In the phenomena which we shall be studying, the luminous vibration may be considered as a scalar quantity. We shall represent it by a function U and shall write (1.1) in the form:

$$\nabla^2 U = \frac{1}{V^2} \frac{\partial^2 U}{\partial t^2} \tag{1.2}$$

For a spherical wave which is a function only of the radius r, and which is emitted from a point source, this equation has a solution:

$$U = \frac{a}{r} \cos(\omega t - \varphi) \tag{1.3}$$

1

in which a is a constant, and a/r is the amplitude of the wave, which decreases in inverse proportion to the distance r from the point source. The factor ω is the angular frequency and φ is the phase. If U depends on only one of the three rectangular coordinates, one has a plane wave, and the propagation equation admits the solution:

$$U = a \cos(\omega t - \varphi) \qquad (1.4)$$

where a is the constant amplitude of the wave.

Figure 1.1 indicates the physical significance of the phase φ. At the point M having the abscissa x, the phase of the vibration at the instant t is the same as it was at the origin O at the time $t - x/V$.

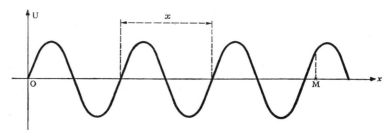

Fig. 1.1

If T is the period, and $\lambda' = VT$ is the wavelength in the medium where the wave is propagating, one has for a plane wave

$$U = a \cos \omega \left(t - \frac{x}{V} \right) = a \cos \left(\omega t - \frac{2\pi x}{VT} \right) = a \cos(\omega t - \varphi) \quad (1.5)$$

with

$$\varphi = \frac{2\pi x}{\lambda'} \qquad (1.6)$$

Let $n = c/V$ be the index of refraction of the medium, where c is the speed of light in a vacuum, and let $\lambda = n\lambda'$ be the wavelength in a vacuum:

$$U = a \cos \left(\omega t - \frac{2\pi n x}{\lambda} \right) \qquad (1.7)$$

The quantity nx, the product of the distance x and the index of refraction n, is called the optical path length (or optical path) between

O and *M*. It is convenient to employ complex notation in the form:

$$U = a[\cos(\omega t - \varphi) + j\sin(\omega t - \varphi)] = ae^{j(\omega t - \varphi)}, \qquad j = \sqrt{-1}$$

$$(1.8)$$

The expression $e^{j\omega t}$ indicates the wave nature of the light and is found as a common factor in all the calculations. One may then omit it from all the calculations and simply represent the vibrations by the expressions:

$$ae^{-j\varphi} \qquad \text{or} \qquad \frac{a}{r}e^{-j\varphi} \qquad (1.9)$$

(plane wave)   (spherical wave)

It is convenient to use complex notation because, if the operations on *U* are linear, one may perform the calculations with the complex functions. The physical quantity is the real part of the final expression given by the calculation. The real amplitude is the modulus of the complex amplitude, and the intensity of the vibration is proportional to the square of the modulus of the complex amplitude. One has

$$I = UU^* \qquad (1.10)$$

A light source such as we are going to consider, which is capable of emitting monochromatic vibrations—that is to say unbounded vibrations, vibrations of infinite duration—is a *temporally coherent* source. This is a theoretical limiting case which does not exist in reality. Light sources emit vibrations of finite duration. The duration of the vibration is called the *coherence time* τ, and the corresponding length $c\tau$, where c is the speed of light, is called the *coherence length*. The coherence length is of the order of 60 cm for the most monochromatic classical sources. It may attain several kilometers for certain lasers.

1.2   The Point Source. Spatial Coherence

A real light source always has finite nonvanishing dimensions, but a particularly important case in optics is that of a "point source." One may define a point source by starting with a function with spherical symmetry, constant on the surface of a sphere of radius ε, and zero everywhere else. The point source is the limit of such a spherical source as ε tends towards zero, while the constant varies in inverse proportion to the surface area of the sphere. Note the similarity between such a point source and the Dirac distribution or delta function (Appendix A, Section A.11).

In all those experiments where one makes use of sources which behave like point sources, one says that there is spatial coherence. This is what happens with lasers, which behave as if one had a point source at the focal point of an objective. The emerging beam is practically cylindrical, and all the points of a cross-section of the beam are on the same wavefront; they are thus in phase and perfectly coherent.

1.3   The Vibration Emitted by a Point Source as Received in a Plane Situated at a Distance *D* from the Source

Consider a point source M, located in the plane $C\eta\zeta$ and having the coordinates η and ζ (Fig. 1.2). It illuminates the parallel plane Fyz situated at a distance D. We wish to calculate the amplitude of the vibration which is emitted by M and which is received at an arbitrary point P of the plane Fyz. According to (1.9) this amplitude may be written:

$$\frac{1}{MP}e^{-jK\cdot MP}, \qquad \varphi = \frac{2\pi}{\lambda}MP = K\cdot MP \tag{1.11}$$

because M emits a spherical wave whose amplitude varies inversely with the distance. Calculate MP assuming that CM and FP are small

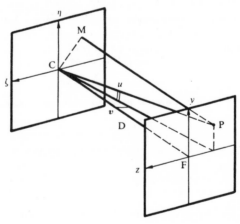

Fig. 1.2

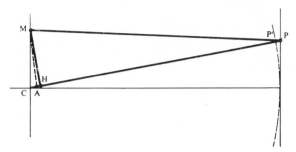

Fig. 1.3

compared to D. Then one has:

$$MP = \sqrt{D^2 + (y - \eta)^2 + (z - \zeta)^2} \cong D + \frac{(y - \eta)^2 + (z - \zeta)^2}{2D}$$

(1.12)

$$MP \cong D + \frac{y^2 + z^2}{2D} + \frac{\eta^2 + \zeta^2}{2D} - \frac{y\eta + z\zeta}{D}$$

(1.13)

These terms represent lengths which may be seen more readily in Fig. 1.3. This is a simplified diagram in which we have taken CM, F, and P in the same plane. The comments which follow apply equally well to Fig. 1.2 and to Fig. 1.3. Draw two spheres, one with center at C and radius CF, the other with center at P and radius PM. The first cuts MP at P' and the second cuts CP at A. If H is the foot of the perpendicular dropped from M on to CP, it is easy to see that, within the range of validity of the approximations we are using,

$$AH = \frac{\eta^2 + \zeta^2}{2D}, \qquad PP' = \frac{y^2 + z^2}{2D}$$

(1.14)

Let $\mathbf{q}$ be a unit vector in the direction of CP; then the projection CH of CM on CP is given by the scalar product:

$$CH = \mathbf{q} \cdot \mathbf{CM} = \frac{y\eta + z\zeta}{D}$$

(1.15)

In all the work that follows, we shall assume that P is sufficiently distant so that for all practical purposes A and H may be taken to coincide:

$$MP \cong D + \frac{y^2 + z^2}{2D} - \frac{y\eta + z\zeta}{D} \qquad (1.16)$$

$$MP' \cong D - \frac{y\eta + z\zeta}{D} \qquad (1.17)$$

Set $u = y/D$ and $v = z/D$; then from (1.11) the amplitude received at P from the source M is:

$$\frac{e^{-jK \cdot D}}{D} \cdot e^{-jK(y^2 + z^2)/2D} \cdot e^{jK(u\eta + v\zeta)} \qquad (1.18)$$

since one may replace MP by D in the denominator, which is a slowly varying function, though not in the exponential, which is swiftly varying. Moreover, if one omits the factor $e^{-jK \cdot D}/D$, which is constant over the plane F, the amplitude received at P from M may be written:

$$e^{jK(u\eta + v\zeta)} \qquad (1.19)$$

Now, if one translates a delta function from C to M, its Fourier transform (appendix A, Section 11) is given by the expression:

$$e^{j \cdot 2\pi(\mu\eta + v\zeta)} \qquad (1.20)$$

and, setting $\mu = u/\lambda$ and $v = v/\lambda$ one sees that the expression (1.19) is the same as the Fourier transform of a delta function. *Consequently, it is not on the plane Fyz, but on the sphere centered at C and of radius CF, that the amplitude emitted by the point source M is identified with the Fourier transform of a delta function.*

1.4   The Vibration Emitted by a Point Source as Received in the Focal Plane of a Lens

Let us investigate the amplitude emitted by a point source M and received at an arbitrary point P in the focal plane of the objective O (Fig. 1.4). Figure 1.5, which is associated with Fig. 1.4, is a simplified drawing like Fig. 1.3 but, of course, the point P may be outside the plane formed by the three points C, M, and F. In accordance with Section 1.3 and to within a constant, the vibration emitted by M and received at P is $e^{-jK(MIP)}$, where the parenthesis represents the optical path. Draw the sphere Σ with center C' and tangent at F to the focal

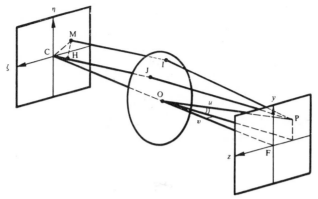

Fig. 1.4

plane Fyz. The point C' is the image of C, and if d is the distance from C to O, the elementary laws of geometrical optics show that $\overline{FC'} = f^2/(d - f)$, where f is the focal length of the objective O. Take CJ parallel to MI. The two rays intersect at P. The ray JP intersects the sphere Σ at P' whose distance from P is given by

$$\overline{PP'} = \frac{y^2 + z^2}{2R}, \qquad R = \frac{f^2}{d - f} \tag{1.21}$$

Malus's theorem allows us to write

$$(MIP) = (HJP) = (HJP') - \overline{PP'} \tag{1.22}$$

But

$$(HJP') = (CJP') - \overline{CH} \tag{1.23}$$

and

$$(CJP') = (COF) = m = \text{constant} \tag{1.24}$$

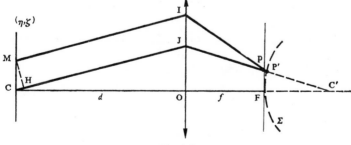

Fig. 1.5

Substituting (1.23) and (1.24) into (1.22) one finds:

$$(MIP) = m - \overline{PP'} - \overline{CH} \tag{1.25}$$

From (1.15) and (1.21) and to within a constant factor e^{-jKm} the vibration emitted by M and received at P may be written:

$$e^{jK(u^2 + z^2)/2R} \cdot e^{jK(u\eta + v\zeta)} \tag{1.26}$$

Since $(MIP') = m - \overline{CH} = m - (u\eta + v\zeta)$, the amplitude emitted by M and received at P' may, to within the same constant factor e^{-jKm}, be written as

$$e^{jK(u\eta + v\zeta)} \tag{1.27}$$

Thus it is on the sphere Σ that the vibration emitted by M and received at P' may be represented by the Fourier transform of a delta function. But if $d \to f$, $R \to \infty$ and $PP' \to 0$:

The vibration produced in the image focal plane of an objective may be represented by the Fourier transform of a delta function if the source is in the object focal plane.

1.5   Important Comments on the Use of the Preceding Formulas

The formulas of the two preceding paragraphs play a fundamental role in the chapters which follow. The phenomena of interference and diffraction always involve the calculation of sums of complex amplitudes at a point P having coordinates y and z. The variables are η and ζ, and those exponentials containing only y and z, as for example the exponential involving $(y^2 + z^2)/2D$ of formula (1.18) or the exponential involving $(y^2 + z^2)/2R$ of formula (1.26), appear only as a factor. Thus if one needs only to calculate the intensity at P, these factors disappear when one multiplies the resultant amplitude by the complex conjugate quantity. To simplify, we shall disregard the exponentials involving $y^2 + z^2$, which are phase terms, and we shall say that *the amplitude produced by a point source M at a distant point or in the focal plane of a lens is given by the Fourier transform of a delta function.*

This simplification is no longer possible if one is interested in the phases.

CHAPTER 2

Interference

2.1 Principle of Interference. Coherent Vibrations

Consider two point sources of light emitting vibrations that are parallel, monochromatic, and unbounded. Since the vibrations are parallel, it is not necessary to consider their direction; the vibrations which may then be regarded as scalar quantities. If two sources of this kind illuminate a plane, one may theoretically expect to find two kinds of phenomena, depending on whether the two vibrations have different frequencies or the same frequency. In the first case, the phenomenon of optical beats is produced. They can be detected only if the frequencies are very nearly the same, and thus one will typically observe the beats in the frequency range of radio waves.

If the frequencies are equal, then—since the velocities of propagation are the same—the phase difference between the two unbounded waves at a given point in space is unchanged with the passage of time; it depends only on the point considered. In certain regions of the plane of observation, the vibrations are in phase and the amplitudes add, i.e., interfere constructively. At these points, there is a maximum amount of illumination. In other regions, the vibrations are in opposition and the amplitudes subtract, i.e., interfere destructively. At these points, there is a minimum amount of illumination. One says that two monochromatic sources which emit parallel vibrations of the same frequency are coherent and that they give rise to *interference phenomena*.

9

Suppose that the two sources are in phase. The two vibrations that originate at the two sources, and that reach an arbitrary point of the plane of observation, have, in general, traversed different optical paths. The difference Δ between the two optical paths is called the *optical path difference*. Those regions of the plane of observation where Δ is equal to an integer multiple of λ, and where consequently there is a maximum amount of light, are called *bright fringes*. Those regions where Δ is equal to an odd number of half-wavelengths correspond to a minimum of light, and are called *dark fringes*. For bright fringes we have, if p is an arbitrary integer:

$$\Delta = p\lambda \tag{2.1}$$

and for dark fringes:

$$\Delta = (2p + 1)\frac{\lambda}{2} = \left(p + \frac{1}{2}\right)\lambda \tag{2.2}$$

One goes from one fringe to the next fringe of the same type when p increases or decreases by unity. The ratio Δ/λ is called the *order of interference*. Bright fringes correspond to integer values of the order of interference, and dark fringes to half an odd integer. If $I_{\max}$ is the intensity of the bright fringes and $I_{\min}$ is the intensity of the dark fringes, one defines the visibility or the contrast of the fringes by the expression:

$$\gamma = \frac{I_{\max} - I_{\min}}{I_{\max} + I_{\min}} \tag{2.3}$$

The contrast is a maximum and equal to unity if the minima are zero, that is to say, if the dark fringes are black fringes. The contrast is zero and there are no longer any visible fringes if $I_{\max} = I_{\min}$.

2.2 Experimental Conditions Needed to Ensure That Interference Phenomena Will Be Observable

The ideal sources which we have just considered emit parallel vibrations of infinite duration. These are temporally coherent sources. We have said that these sources are mutually coherent and capable of interference because, at an arbitrary point in space, the phase difference between the two vibrations emitted is independent of the time, since their vibrations have the same frequency and the same speed of propagation. Each source therefore emits "a vibration" and the observation is made within the "duration" of the vibration. Since this is infinite, one has at one's disposal all the time needed to make an observation. Lasers do not give rise to vibrations that exist for an infinite period, but rather vibrations that last long enough so that it is possible to observe interference phenomena during the time of emission of a single burst of radiation from a laser. Similarly, the vibrations of a tuning fork last long enough to permit one to hear many beats before it becomes necessary to re-excite it.

With thermal light sources, things are very different, because the coherence time is so short that one receives a considerable number of independent vibrations during the period of time needed to make an observation. The receiver can respond only to an average of the effects produced by the rapid succession of different phenomena which are all mixed together as they arrive at the receiver. Two thermal point sources are incapable of giving rise to observable interference effects; one says that they are incoherent. From these considerations it follows that in order to observe interference with thermal sources, it is essential that the two vibrations which are to be superimposed must proceed from the same light source. Rays originating in a single point source are superimposed after having traversed different optical paths. An interference apparatus thus appears as a wavefront splitter; it splits the incident wave into two or more waves which, after having traversed different paths, are recombined and thus give rise to interference phenomena.

As we remarked at the beginning of this section, the two sources emit parallel vibrations. In fact, two perpendicular vibrations have as their resultant an elliptic vibration whose energy is independent of the path difference of the component vibrations. The resultant energy is simply equal to the sum of the energies of the component vibrations. It is necessary to state explicitly that two nonparallel vibrations A and B may give rise to interference phenomena because one of the two vibrations, A, for example, may be decomposed into two components A_1 and A_2, A_1 being parallel to B and A_2 being perpendicular to B. The vibrations A_1 and B give rise to interference phenomena. The vibrations A_2 and B do not interfere. The vibration A_2 adds its intensity to the intensity variations resulting from the interference between A_1 and B, and reduces the contrast of the fringes.

The diagrams which follow, in which G is a plate having plane-parallel semireflecting faces, summarize the preceding discussion.

Thermal light source

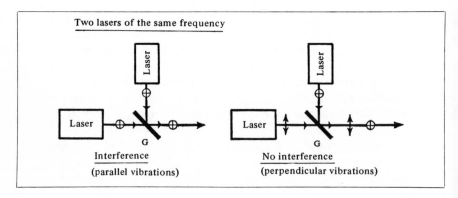

Interference observable with a single source
(the plate G is a wave-divider)

Two lasers of the same frequency

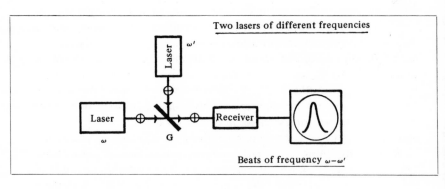

Interference   No interference
(parallel vibrations)   (perpendicular vibrations)

Two lasers of different frequencies

Beats of frequency $\omega - \omega'$

In principle, two lasers can interfere, but the experiments are very difficult to carry out; and *in all cases one actually uses a single source and a wave-divider.*

With a single source, it is unnecessary to display the direction of the vibrations if the relative inclination ε of the interfering vibrations (Fig. 2.1) is zero or very small, a condition which we shall assume is always satisfied. Besides, if the orientation and the phase of the vibrations emitted by the source do vary with the passage of time, these variations will affect all the vibrations that interfere at *P* in the same

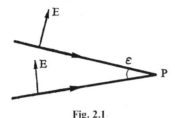

Fig. 2.1

way. Of course, it is necessary either that the wavedivider is not itself affected by these changes or that it transmit them in the same fashion to the two vibrations. Finally, one may say that interference phenomena are observable:

(a)   if the vibrations are parallel

(b)   with distinct sources of the same frequency if the temporal coherence is sufficiently great (lasers), but the experiment is very difficult, and

(c)   With a single source and a wavedivider if the temporal coherence is otherwise insufficient (thermal light sources). This is the method actually used in all cases.

2.3   Nature of the Coherent Sources Used to Give Rise to Interference Phenomena

We shall consider the ideal case of point sources emitting monochromatic parallel vibrations of the same frequency. It is well known that such sources do not exist in reality and that instead it is necessary to use a single light source and a wavefront splitter. It is not necessary at the moment to get too involved in the details of the way in which the coherent sources are produced by means of wavefront division.

Achieving this experimentally requires us to consider diffraction phenomena, which will be studied in Chapter 3. We shall speak of coherent point sources in order to distinguish them from real light sources. The phenomena produced by two coherent point sources are called "two-beam interference." Interference phenomena produced by a very large number of coherent point sources are called "multiple-beam interference."

2.4   Localization of Two-Beam Interference Fringes

Let S be a monochromatic point light source (Fig. 2.2). In the figure the beam-splitter consists of four mirrors M_1, M_2, M_3, and M_4. Of course, this is only an example and there are many other types of experimental arrangement. The two rays SI_1J_1P and SI_2J_2P traverse two different paths before being superimposed at an arbitrary point P where they interfere.

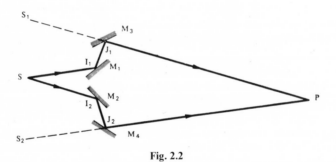

Fig. 2.2

The path difference Δ of the vibrations that arrive at P is well defined. So also is the resultant amplitude, and consequently also the intensity. Since S is a point source, no other vibration will arrive at P that will be capable of causing confusion in the interpretation of the phenomena. But P is an arbitrary point, and what we have just said is valid in any region whatever of space where the vibrations are superimposed. At each point of space there are two vibrations which interfere without any confusion caused by other vibrations. Interference phenomena are observable everywhere in space where the two vibrations are superimposed. *A point source gives rise to interference phenomena which are not localized.*

The virtual images S_1 and S_2 of the source S are formed respectively by the pair of mirrors M_1 and M_3 and the pair of mirrors M_2 and M_4. These images S_1 and S_2 behave like two *coherent point sources*, and one may very reasonably consider that the vibrations which interfere at P originate in S_1 and S_2 without it being necessary to speak of S at all. This is what we shall do in this chapter. Since the sources S_1 and S_2 originate in a point light source, the interference phenomena which we shall describe are nonlocalized.

In some later work, we shall study two-beam interference caused by light originating in extended sources (Chapter 8). We shall see that in this case the fringes are localised in a well-defined region of space.

2.5   Interference Phenomena Produced by Two Coherent Point Sources. Young's Interference Fringes

We shall consider two coherent point sources emitting parallel monochromatic vibrations of the same frequency and phase. These sources produce two-beam interference phenomena. The two sources M_1 and M_2 are located on the axis $C\zeta$ (Fig. 2.3), they are symmetrically placed with respect to C, and their separation is $M_1 M_2 = \zeta_0$. Since $M_1 F = M_2 F$ the vibrations emitted in phase by M_1 and M_2 are also in phase at the point F. We shall consider the phenomena at a point $P(y, z)$ in the plane Fyz, which is parallel to the $\zeta\eta$ plane, and which is at a distance D from the sources M_1 and M_2. We shall assume that $M_1 M_2$ and FP are small compared to D. In accordance with paragraph 1.5 and formula (1.18), the amplitudes emitted at M_1 and M_2 and

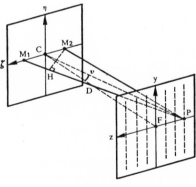

Fig. 2.3

received at P are given by the Fourier transforms of the two delta functions corresponding to M_1 and M_2. The resultant amplitude at P is:

$$f(v) = e^{j(Kv\zeta_0/2)} + e^{-j(Kv\zeta_0/2)} = 2\cos\frac{Kv\zeta_0}{2} \qquad (2.4)$$

from which the intensity is

$$I = I_0 \cos^2\frac{Kv\zeta_0}{2} \qquad (2.5)$$

I_0 being a constant. In order to be able to see things more physically, it will suffice to return to formula (1.11). The amplitudes emitted by M_1 and M_2 and received at P may be written:

$$\frac{e^{-jK\overline{M_1P}}}{M_1P}, \quad \frac{e^{-jK\overline{M_2P}}}{M_2P} \qquad (2.6)$$

In the denominators we may replace M_1P and M_2P by D and we may then omit this factor, since it is constant. The amplitude at P is then:

$$f(v) = e^{-jK\overline{M_1P}} + e^{-jK\overline{M_2P}}$$
$$= e^{-jK\overline{M_1P}}[1 + e^{jK(\overline{M_1P}-\overline{M_2P})}] \qquad (2.7)$$

$\overline{M_1P} - \overline{M_2P}$ represents the path difference Δ between the vibrations arriving at P from M_1 and M_2. Projecting M_1M_2 on M_1P as was done in Section 1.3, we find $\Delta = v\zeta_0$ and the formula (2.5) takes on the general form:

$$I = I_0 \cos^2\frac{K\Delta}{2} \qquad (2.8)$$

To the approximation we are using, the path difference Δ does not vary when the point P is displaced parallel to Fy. The fringes are therefore parallel straight lines, which are called Young's interference fringes. These fringes have contrast equal to unity (Section 2.1). If $\Delta = p\lambda$ one has the pth bright fringe, and if $\Delta = (p + \frac{1}{2})\lambda$ one has the pth dark fringe. If one allows p to increase by unity, formula (2.5) shows that the distance separating two consecutive bright fringes or two consecutive dark fringes is, in angular measure, $v = \lambda/\zeta_0$, or, in linear measure, $vD = \lambda D/\zeta_0$ (Fig. 2.4). The spacing of the fringes varies with the wavelength.

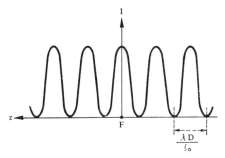

Fig. 2.4

When the amplitudes emitted by M_1 and M_2 are unequal, the amplitude may be written using (2.4) or (2.7) as:

$$f = a + b e^{jK\Delta} \qquad (2.9)$$

From this, setting $\varphi = K\Delta = 2\pi\Delta/\lambda$, the intensity may be written as

$$I = ff^* = a^2 + b^2 + 2ab \cos\varphi \qquad (2.10)$$

which is the classical Fresnel formula, and which may also be obtained from (2.4) or (2.7). The intensity of the bright fringes is $(a + b)^2$, and the intensity of the dark fringes is $(a - b)^2$, from which the contrast is

$$\gamma = \frac{2ab}{a^2 + b^2} \qquad (2.11)$$

The contrast is a maximum when the amplitudes of the interfering vibrations are equal.

Analogous phenomena are observed when a lens is used (Fig. 2.5). The plane $C\eta\zeta$ containing the two sources M_1 and M_2 is at an arbitrary distance from the lens O and one makes the observations in the image focal plane. According to our earlier discussion, one may say that the amplitude in the plane Fyz is once again given by the Fourier transform of two delta functions corresponding to M_1 and M_2. These transforms are $e^{j(Kv\zeta_0/2)}$ and $e^{-j(Kv\zeta_0/2)}$, and we come again to the same results.

It is necessary to remark that in the experiment sketched in Fig. 2.3, the distance is not predetermined, and that consequently the fringes are not localized. Similarly, if one uses the setup of Fig. 2.5; under the conditions shown there, fringes may be observed not only in the focal plane, but also outside this plane. We need only ponder the fact that

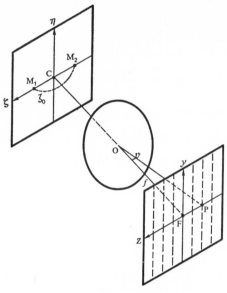

Fig. 2.5

the images of M_1 and M_2 formed by the lens O directly illuminate the plane Fyz in order to see that this reduces to the preceding case. This results immediately from the discussion of Section 2.4. The reason for considering the phenomena in the focal plane of a lens will become apparent in the following chapter.

2.6   Interference Phenomena Produced by an Infinite Number of Coherent, Equidistant, Point Sources (Grating)

Let us consider coherent sources emitting monochromatic parallel vibrations, all of the same frequency. They are situated in the plane $C\eta\zeta$ (Fig. 2.6) along the axis $C\zeta$. We shall suppose them to be equidistantly spaced and in phase. We shall observe the phenomena in the focal plane of a lens O. The grating formed by an ensemble of coherent point sources, that is, by an infinity of delta functions, constitutes a *Dirac comb* (Appendix A, Section A.14). The "step" of the grating is equal to the distance ζ_0 separating two consecutive sources. At the focal point of the objective O, we shall still say that the amplitude is given by the Fourier transform of an infinity of delta functions, that is to say,

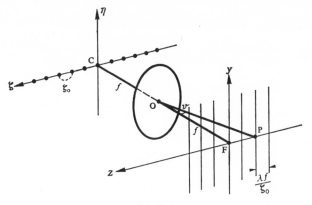

Fig. 2.6

by the Fourier transform of a Dirac comb. But the transform of a Dirac comb of step ζ_0 is another Dirac comb of step $\lambda f/\zeta_0$, where f is the focal length of the objective. One has (Appendix A, Section A.14), setting $v_0 = \lambda/\zeta_0$:

$$\text{F.T.}\left[\text{comb}\left(\frac{\zeta}{\zeta_0}\right)\right] = \sum_{n=-\infty}^{+\infty} \delta\left(v - n\frac{\zeta}{\zeta_0}\right) = \text{comb}\left(\frac{v}{v_0}\right) \quad (2.12)$$

because, in the relation (A.36) we have seen that it is necessary to set $v = v/\lambda$ (Section 1.3). The angular step is $v_0 = \lambda/\zeta_0$ and the linear step is $fv = \lambda f/\zeta_0$. The Fourier transform depends only on v, and consequently the amplitude is unchanged when one moves along an arbitrary line parallel to Fy. In the focal plane of the objective O, we observe very narrow interference fringes consisting of bright lines parallel to Fy and separated by dark spaces. These are the *spectra* of the grating, and their separation $\lambda f/\zeta_0$ varies with the wavelength λ.

For any particular one of these spectra, say for the pth spectrum measured from the center F, the vibrations emitted by all the coherent point sources arrive in phase at P. Indeed, in an arbitrary direction the path difference Δ between the vibrations emitted by two consecutive sources is $v\zeta_0$ in the direction v (Fig. 2.7). For the pth spectrum:

$$v = p\frac{\lambda}{\zeta_0} \quad (2.13)$$

and the path difference is $\Delta = v\zeta_0 = p\lambda$. Outside of these directions, the phase differences $2\pi\Delta/\lambda$ take on all possible values between 0 and

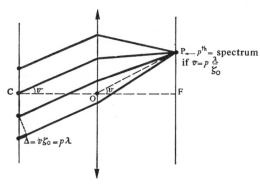

Fig. 2.7

2π. One may always associate with a particular vibration another vibration which is in opposition to the first. These vibrations annul each other in pairs, and there is no illumination between the spectral lines. To the approximation employed, the path difference does not change if the point of observation P (Fig. 2.6) moves parallel to Fy and the spectra are in fact "spectral lines" parallel to Fy as we stated earlier. Figure 2.7 shows that the path difference Δ is the same for all the ensemble as long as we restrict ourselves to parallel rays. It is only in the focal plane that one can observe the spectra. The multiple-beam fringes observed in this experiment are therefore localized at infinity or in the focal plane of a lens.

2.7   Remark on the Structure of the Fringes

The interference fringes produced by two coherent point sources are characterized as far as intensity is concerned by a $\cos^2 x$ law (2.7). The fringes produced by a very large number of sources have a very different structure from that of the two-beam fringes; these fringes are in fact very much sharper, as was shown in the results of Section 2.6.

2.8   White Light Interference Phenomena

To begin with, let us consider again the experiments shown in Figs. 2.3 or 2.5. In white light, the intensities of the interference patterns corresponding to different wavelengths are superposed. If the two coherent point sources M_1 and M_2 are in phase, at the center F of the interference pattern $(M_1F = M_2F)$ one has a maximum of intensity regardless of the wavelength; one observes a white central fringe. But as soon as one moves away from the center, the fringes corresponding to different wavelengths no longer coincide. The red fringes are wider than the white fringes. At an arbitrary point P of the plane of observation, it may happen that for a single wavelength one has $\Delta/\lambda = p + \frac{1}{2}$. In the neighborhood of this point, the vibration of wavelength λ is absent and the superposition of the other radiations no longer gives white; a color appears. One thus observes colors on either side of the white central fringe. These are the interference tints with a white center. In other cases which we shall study, the central fringe is black and the colors are described as having a black center. These two color scales, called Newton's scales, are given in tables which permit a given color to be easily referenced. In proportion as one moves away from the central fringe, Δ increases, and it may happen that at some particular point one has:

$$\frac{\Delta}{\lambda_1} = p_1 + \frac{1}{2}, \qquad \frac{\Delta}{\lambda_2} = p_2 + \frac{1}{2}, \qquad p_1 \neq p_2 \qquad (2.14)$$

therefore two radiations are missing simultaneously at the point P. One may note that the color is less vivid than in the case where a single radiation is extinguished. Moving still further away from the central fringe, the number of radiations extinguished at a given point continues to increase. The colors become less and less vivid, and when four or five wavelengths are missing at the same point, one again encounters a white which is called a white of a higher order.

Instead of observing the colored fringes in the plane Fyz (Fig. 2.5), we may place the slit of a spectrograph parallel to the fringes and observe the spectra. To each absent wavelength there corresponds a black "channel." The distribution of intensities in the spectrum as a function of wavelength is given by (2.7) with $v\zeta_0 = $ constant. Suppose that λ_1 and λ_2 correspond to the ends of the visible spectrum. From (2.14) the number of black channels is given by:

$$p_2 - p_1 = \Delta\left(\frac{1}{\lambda_2} - \frac{1}{\lambda_1}\right), \qquad \lambda_2 < \lambda_1 \qquad (2.15)$$

The number of channels is given by the difference of the orders of interference corresponding to the extremes of the visible spectrum, since p increases or decreases by one unit whenever one goes from one channel to the next one.

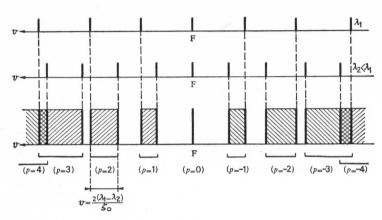

Fig. 2.8

Consider now the case of the infinite grating of Figs. 2.6 and 2.7. If the coherent point sources emit monochromatic light of wavelength λ_1, the spectra constitute a Dirac comb of angular step λ_1/ζ_0 (Fig. 2.8). If the sources emit monochromatic light of wavelength λ_2, the spectra form a Dirac comb having a different step λ_2/ζ_0. In Fig. 2.8 the spectra are indicated by vertical lines, and in the drawing at the bottom one has the structure of the spectra, assuming that the coherent point sources emit simultaneously the two radiations of wavelengths λ_1 and λ_2. Now suppose that the light emitted is white light, and that the wavelengths λ_1 and λ_2 correspond to the extremes of the visible spectrum. One then has an infinite number of Dirac combs, and the intervals on the wavelength scale enclosed within the spectra due to λ_1 and λ_2 are filled with light. These intervals are indicated by brackets in Fig. 2.8. These are still called spectra: spectra of the first order for $p = \pm 1$, of the second order for $p = \pm 2$, etc. They are evidently symmetric with respect to F ($p = 0$), where we find not a spectrum but a white "spectral line." The path difference Δ is in fact zero at F, regardless of the wavelength. In accord with (2.13) we see that the angular width of the pth spectrum is

$$v = p\frac{\lambda_1 - \lambda_2}{\zeta_0} \tag{2.16}$$

This spreading out of the spectra is called the dispersion of the grating. The dispersion increases with the order of the spectrum. In the general case of an arbitrary spectrum, formula (2.16) shows that the position of a radiation of a given wavelength varies linearly as a function of v; one says that *the spectrum is normal.*

In a particular direction v, there may be superposition of a spectral line of wavelength λ_1 corresponding to the spectrum of order p, and a spectral line of wavelength λ_2 corresponding to the spectrum of order $p \pm 1$. In Fig. 2.8 there is a partial superposition of the spectra of orders 3 and 4. The regions of superposition may be calculated with the help of formula (2.16), the wavelengths λ_1 and λ_2 being inside or outside the visible spectrum.

2.9   Stationary Waves

In the phenomena which we have just studied, the waves which interfered were propagating in the same direction and in the same sense. It is possible to observe interference of two waves which are propagating in opposite senses. Consider, for example, a bundle of parallel rays (plane wave Σ) normal to the mirror M (Fig. 2.9). At an arbitrary point

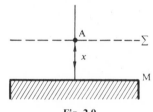

Fig. 2.9

A located at a distance x from the mirror there will be interference between an incident wave and a wave which has been reflected by M and is retracing its path in the opposite sense. The geometrical path difference is $\Delta = 2x$. The reflection itself at M may introduce a change of phase and the path difference is then written $\Delta = 2x + \varepsilon$. Suppose that the reflectivity of the mirror M is sufficiently high that we may consider the amplitudes of the incident wave and the reflected wave to be practically equal. In accordance with (2.8) the intensity at A is:

$$I = I_0 \cos^2 \frac{\pi \Delta}{\lambda} = I_0 \cos^2 \frac{\pi(2x + \varepsilon)}{\lambda} \qquad (2.17)$$

There is a maximum of illumination if:

$$\frac{\Delta}{\lambda} = p, \qquad x = p\frac{\lambda}{2} + \frac{\varepsilon}{2} \qquad (2.18)$$

These planes, located at a distance x from the mirror satisfying (2.18), are the ventral or antinodal planes. The distance between two consecutive antinodal planes is equal to $\lambda/2$. There is a minimum of illumination if:

$$\frac{\Delta}{\lambda} = p + \frac{1}{2}, \qquad x = \left(p + \frac{1}{2}\right)\frac{\lambda}{2} + \frac{\varepsilon}{2} \qquad (2.19)$$

The planes defined by (2.19) are the nodal planes. The distance between two consecutive nodal planes is also equal to $\lambda/2$.

Wiener was able to photograph this phenomenon by using a very thin sensitive film deposited on a plate of glass AB (Fig. 2.10). The plate AB made a very small angle with the mirror M. It cut the antinodal planes in a system of straight lines parallel to the line of intersection A at the vertex of the angle formed by the two planes. After development one observes a set of fringes corresponding to the intersections with the antinodal planes.

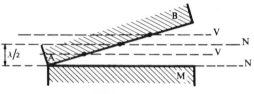

Fig. 2.10

In order to find out whether reflection at a glass surface occurs with or without a change of phase, one applies the sensitive film to a spherical surface M *made out of glass* (Fig. 2.11). The lower surface of M is blackened in order to prevent all undesired reflection. The photograph shows a system of annular rings so disposed that the central portion corresponds to a minimum. There is thus a node at the contact with M at the point O, and the reflection at the glass surface introduces a change of phase equal to π.

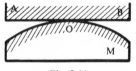

Fig. 2.11

Electromagnetic theory indicates that it is the electric field which has a node at M. It is thus the electric field which is responsible for the photochemical action of the light wave in this experiment.

CHAPTER 3

Diffraction

3.1 The Existence of Diffraction Phenomena

Let S be a point source which illuminates a screen E_2 through the aperture T in the screen E_1 (Fig. 3.1). Instead of observing on E_2 "the shadow" of the aperture T, one finds that the light spreads out more than is predicted by geometrical optics. One says that the aperture T *diffracts* the light. The smaller the aperture T, the more the diffracted light spreads out over the screen E_2. The diffraction phenomena observed on the screen E_2 are called *diffraction at a finite distance* or *Fresnel diffraction*.

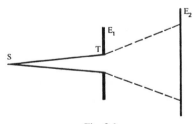

Fig. 3.1

Except for the last paragraph of this chapter, we shall be studying the diffraction phenomena when *the source S and the screen E_2 are at infinity*. Thus one is dealing with the experimental setup shown in Fig. 3.2. The source S is at the focal point of a collimator objective O_1,

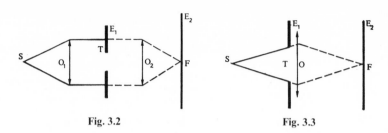

Fig. 3.2   Fig. 3.3

the aperture T is illuminated by a plane wave, and one observes the phenomena in the image focal plane E_2 of a second objective O_2. The diffraction phenomena observed under these conditions are described as phenomena of *diffraction at infinity* or *Fraunhofer diffraction phenomena*. We may remark that the two objectives O_1 and O_2 may be replaced by a single objective O (Fig. 3.3), the screen E_1 pierced by the aperture T being placed in contact with the objective. Moreover, the aperture may be bounded by the rim of the objective O itself.

3.2   Principle of Huygens and Fresnel

Consider a point source S (Fig. 3.4) and let Σ be the wave front at time t. According to Huygens's principle, each point M of the wavefront Σ is considered as a secondary source which emits a spherical wavelet in a homogeneous isotropic medium. At the time $t + \theta$ the radius of the wavelet is $V\theta$, V being the velocity of propagation. At the time $t + \theta$ the wave Σ' is the envelope of the wavelets of radius $V\theta$, and thus is a sphere of radius $V(t + \theta)$. Huygens thus made evident the mechanism of the propagation of the phenomenon as a step-by-step process from one point of space to another. Huygens' construction was brought to completion by Fresnel's hypothesis, according to which there could be interference among the different wavelets. It thus would seem very natural to assume that the secondary sources distributed over Σ would have precisely the phase corresponding to the state of the vibration over this wave surface. A more complete study of these phenomena (see Appendix B) shows that it is in fact necessary to introduce an advance of phase equal to $\pi/2$.

The principle of Huygens–Fresnel not only predicts an amplitude in the direction of propagation, but also predicts an amplitude in the opposite sense to the direction of propagation produced by a backward-moving wave Σ'' which is the other envelope of the wavelets (Fig. 3.5). A

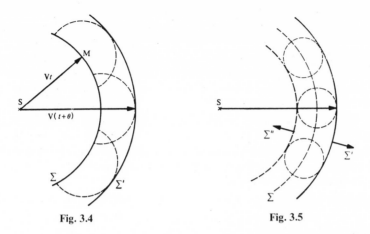

Fig. 3.4   **Fig. 3.5**

mathematical study of these phenomena has made it possible to justify
the Huygens–Fresnel principle, and to explain the fact that the wave
Σ'', which is contrary to experience, actually does not appear.

The Huygens–Fresnel principle allows us to calculate in a simple
fashion the phenomena of diffraction at infinity which we shall consider.
Let O be an objective illuminated by a parallel bundle of rays (Fig. 3.6).
The incident plane wave is bounded by a screen E_1 pierced by an
aperture T, and the geometric image of the source is located at F.
According to the Huygens–Fresnel principle, all the elements of Σ
behave like secondary sources in phase. In Fig. 3.6, we have shown
a group of rays diffracted in the direction v. These rays are brought to
a focus at the point P in the focal plane E_2 of the objective O. The vibration at P is thus the resultant of an infinite number of vibrations emitted

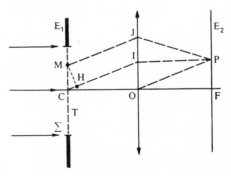

Fig. 3.6

by the secondary sources distributed over Σ. The calculation of the
state of vibration at *P* thus reduces to an interference calculation.
We shall assume that the aperture of the objective *O* is small; under
these circumstances, the rays which arrive at *P* have very small inclinations one with respect to the other, and the vibrations are
practically parallel. It is therefore not necessary to take the direction
of the vibrations explicitly into account; the vibrations may then be
considered as if they were scalar quantities. We shall assume, moreover,
that the amplitude of the radiation emitted by the secondary sources is
independent of the direction, that is, of the angle *v*. This approximation
is valid as long as *v* is small.

3.3   Simplified Expression for the Amplitude at an Arbitrary
Point in the Plane of Observation. Diffraction at Infinity

To begin with, let us state precisely the conditions in which one
can observe Fresnel and Fraunhofer diffraction phenomena. Let *T*
be an aperture illuminated by a point source at infinity (Fig. 3.7). We
shall assume that the plane wavefront Σ coincides with E_1. We wish
to calculate the amplitude diffracted to an arbitrary point *P* of the
screen E_2 by the aperture *T*. According to Huygens' principle, an arbitrary point *M* of the wave Σ gives rise at *P* to the amplitude given by
the expression (1.11). Using (1.12), the amplitude at *P* is:

$$\frac{e^{-jK \cdot D}}{D} e^{-j(K/2D)[(y-\eta)^2+(z-\zeta)^2]} \tag{3.1}$$

from which the total amplitude received at *P* from the aperture *T* is:

$$f(y, z) = \frac{e^{-jK \cdot D}}{D} \iint_T e^{-j(K/2D)[(y-\eta)^2+(z-\eta)^2]} \, d\eta \, d\zeta \tag{3.2}$$

Studying the amplitude at *P* by means of the integral (3.2) amounts to
studying the phenomenon of diffraction at a finite distance, in other
words Fresnel diffraction. In this calculation, we assume that *CM*
and *FP* remain small with respect to *D*. If now the point *P* moves
sufficiently far away, one may make an additional approximation and
neglect the term $(\eta^2 + \zeta^2)/2D$ as we did in Sections 1.3 and 1.4. The
amplitude at *P* may then be written:

$$f(y, z) = \frac{e^{-jK \cdot D}}{D} e^{-jK(y^2+z^2)/2D} \iint_T e^{j(K/D)(y\eta+z\zeta)} \, d\eta \, d\zeta \tag{3.3}$$

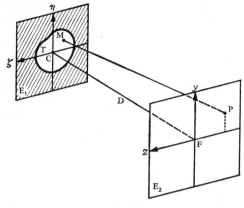

Fig. 3.7

Since we are calculating relative intensities, it is not necessary to retain the factor in front of the integral in (3.3). Setting $u = y/D$ and $v = z/D$, we have:

$$f(u, v) = \iint_T e^{jK(u\eta + v\zeta)} \, d\eta \, d\zeta \qquad (3.4)$$

Recalling the conventions already employed (Section 1.5), we may immediately recover the expression (3.4). Each point M of the wavefront gives rise at P to an amplitude which is the Fourier transform of a delta function. In order to determine the total amplitude due to the

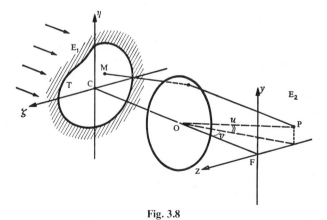

Fig. 3.8

whole aperture, it suffices to sum the individual contributions. The particular conditions required to observe Fraunhofer diffraction phenomena are fulfilled when one makes observations at infinity in accordance with the experimental arrangement shown schematically in Fig. 3.6. From M, drop the perpendicular MH on to CI. In accordance with (1.15), we see that $u\eta + v\zeta$ represents the path difference $CH = (CIP) - (MJP)$. Indeed, Malus's theorem shows that the optical paths (MJP) and (HIP) are equal. Since the point P is not necessarily in the plane of symmetry of the system, Fig. 3.8 represents the situation in perspective.

We may represent the unobstructed part of the wavefront by a function $F(\eta, \zeta)$ which is constant in the interior of T and zero everywhere else. The amplitude at P due to the whole aperture T is, to within a constant factor:

$$f(u, v) = \int\limits_{-\infty}^{+\infty}\!\!\!\int F(\eta, \zeta)e^{jK(u\eta + v\zeta)}\, d\eta\, d\zeta \qquad (3.5)$$

The diffraction phenomena in the plane E_2 are described by the Fourier transform of the distribution $F(\eta, \zeta)$ of amplitudes on the wavefront. The amplitude and phase need not be constant over the wavefront, for example because one might have placed in the aperture a plate of glass of variable absorption and thickness. These two effects may be represented by a complex function $F(\eta, \zeta)$:

$$F(\eta, \zeta) = A(\eta, \zeta)e^{j\varphi(\eta, \zeta)} \qquad (3.6)$$

where $A(\eta, \zeta)$ specifies the amplitude distribution and $\varphi(\eta, \zeta)$ specifies the phase distribution on the wavefront. If $\varphi(\eta, \zeta) = 0$ the wave Σ is perfectly plane and $A(\eta, \zeta)$ gives the amplitude distribution. If $A(\eta, \zeta)$ = a constant with $\varphi(\eta, \zeta) \neq 0$, the wavefront is deformed, but the amplitude is constant at all points. Therefore if $F(\eta, \zeta)$ is real, there are no phase variations over the wavefront.

The luminous intensity at P is given by

$$I = f(u, v)f^*(u, v) \qquad (3.7)$$

Formula (3.5) was derived subject to the assumption that the presence of the screen E_1 did not perturb the wavefront Σ. A rigorous solution, which we do not contemplate, would involve going back to the equation of propagation and taking into account the screen E_1.

In all cases, one assumes that the screen E_1 causes no change in the wave other than the suppression of the parts masked off by it, that is to say, that the electric field (the luminous vibration) at an arbitrary point of Σ is the same with or without the screen E_1. This approximation is valid if the dimensions of the aperture are large compared to the wavelength. Within the applicable range of this hypothesis a more complete calculation than the preceding work gives for the amplitude at P the expression (see Appendix B):

$$f(u, v) = \frac{j}{\lambda f} \int\limits_{-\infty}^{\infty}\!\!\int F(\eta, \zeta)e^{jK(u\eta + v\zeta)} \, d\eta \, d\zeta \qquad (3.8)$$

Particularly note the factor j which gives the correct phase.

In all the diffraction calculations we shall be making, we shall be looking for the intensity distributions, and the factor in front of the integral (3.8) may be omitted.

That is why in the work which follows we shall calculate the amplitudes by means of formula (3.5) which represents the Fourier transform of the complex amplitude distribution over the wavefront Σ'.

From our knowledge of the properties of the Fourier transformation (Appendix A, Section A.10), if $f(u, v)$ is the transform of $F(\eta, \zeta)$, the diffraction phenomenon as described in terms of the intensity $| f(u, v)|^2$ is given by the transform of the autocorrelation function of $F(\eta, \zeta)$. This remark often makes possible a simple calculation of the diffraction pattern as far as intensity is concerned.

3.4 Diffraction Pattern at Infinity Due to a Rectangular Aperture

The aperture, illuminated by a plane wave, is a slit of width ζ' and height η'. We have (Section A.5)

$$F(\eta, \zeta) = \text{Rect}\!\left(\frac{\eta}{\eta'}\right) \text{Rect}\!\left(\frac{\zeta}{\zeta'}\right) = 1, \qquad |\eta| < \frac{\eta'}{2}, \quad |\zeta| < \frac{\zeta'}{2}$$

$$F(\eta, \zeta) = 0, \qquad\qquad\qquad\qquad \text{everywhere else} \qquad (3.9)$$

The amplitude at P is given by the Fourier transform of a two-dimensional rectangle function (Appendix A, Section 5). We have, to within a constant factor:

$$f(u, v) = \text{sinc}\left(\frac{\pi u \eta'}{\lambda}\right)\text{sinc}\left(\frac{\pi v \zeta'}{\lambda}\right) \tag{3.10}$$

The integral (3.5) shows that the constant factor is proportional to the surface area of the aperture. This is a general result, valid no matter what the form of the aperture: the amplitude at a point in a diffraction pattern *is proportional to the surface area of the aperture*. The intensity is *proportional to the square of the surface area of the aperture*.

We observe in the focal plane E_2 two sets of diffraction fringes which form a sort of checkerboard pattern (Fig. 3.9). Along Fz the intensity distribution is given by $\text{sinc}^2(\pi v \zeta'/\lambda)$ and along Fy by $\text{sinc}^2(\pi u \eta'/\lambda)$.

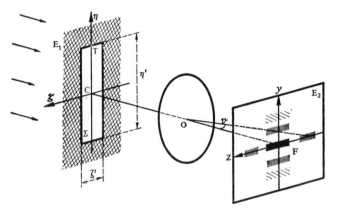

Fig. 3.9

The central rectangle, which extends as far as the first zero ($u = \lambda/\eta'$, $v = \lambda/\zeta'$) is called *the central diffraction spot*. The central spot spreads out further as the aperture becomes smaller. This is a general result which is valid whatever may be the form of the aperture. As soon as one moves away from the central spot, the intensity becomes very small.

3.5   Diffraction at Infinity by a Thin Slit

If, for example, $\eta' \gg \zeta'$, the aperture reduces to a thin slit parallel to $C\eta$. The factor $\text{sinc}(\pi u \eta'/\lambda)$ in expression (3.10) is practically zero everywhere except on Fz; the diffracted light spreads out only along

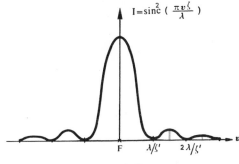

Fig. 3.10

Fz, that is to say perpendicularly to the direction of the slit. The smaller the width of the slit, the more the diffraction pattern spreads out. The intensity variations are represented by the curve in Fig. 3.10. Up till now we have assumed that E_1 is illuminated by a plane wave, that is to say, by a point source at infinity. There is nothing to keep us from replacing the point source by a very thin slit source at infinity, parallel to the slit located in the screen E_1. To each point of the source there corresponds the diffraction phenomena we have just studied. All these phenomena add up in intensity and one observes a perfectly clear diffraction pattern consisting of diffraction fringes parallel to both slits. In the direction perpendicular to these fringes, the variation of intensity is still given by the curve of Fig. 3.10. If the source slit is not parallel to the slit E_1, the diffraction pattern is still sharp, but the diffraction fringes are inclined and are parallel to the source slit.

3.6   Diffraction at Infinity by a Circular Aperture

The aperture, illuminated by a plane wave parallel to the plane of the screen and aperture, is bounded by a circle of radius a_0 (Fig. 3.11). We then have (Appendix A, Section 5):

$$F(\eta, \zeta) = 1, \qquad \eta^2 + \zeta^2 < a_0^2$$
$$F(\eta, \zeta) = 0, \qquad \eta^2 + \zeta^2 > a_0^2 \tag{3.11}$$

The amplitude at P is given by the Fourier transform of a circle function:

$$f(\alpha) = \frac{2J_1(Z)}{Z} \tag{3.12}$$

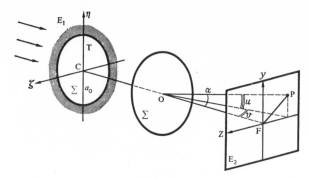

Fig. 3.11

with

$$Z = \frac{2\pi a_0 \alpha}{\lambda} = K a_0 \alpha, \qquad \alpha = \sqrt{u^2 + v^2} \qquad (3.13)$$

where $J_1(Z)$ is the Bessel function of order 1 and argument Z.

The curve in Fig. 3.12 shows the variation of the intensity $I = ff^*$ as a function of Z. The diffraction pattern, or Airy spot, is composed of a very bright central spot surrounded by rings which are alternately dark and bright. The diameter of the central spot increases as the diameter of the aperture decreases. The maxima of illumination in the bright rings are much less intense than the central maximum, and decrease rapidly with the order of the bright ring. For the first black ring, one finds $Z = 3.83$, from which the radius of the first black ring, that is to say the radius of the central diffraction spot, from (3.13), is:

$$\rho = f\alpha = \frac{1.22\lambda}{2(a_0/f)} \qquad (3.14)$$

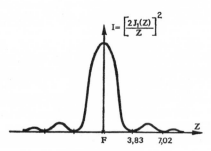

Fig. 3.12

and the angular radius is:

$$\alpha = \frac{1.22\lambda}{2a_0} \tag{3.15}$$

Objectives having the same f/number, $f/2a_0$, but different diameters give the same diffraction spot as far as linear dimensions are concerned.

Let us assume that $F(\eta, \zeta)$ no longer represents a distribution of constant amplitude. Suppose instead that we have an amplitude variation which follows a Gaussian law. This may be accomplished by placing in front of the objective a plane-parallel plate of glass whose transmission function, which is a function of revolution around the axis, obeys the Gaussian law as far as amplitude is concerned. Since the Fourier transform of a Gaussian function is again a Gaussian function, the profile of the diffraction pattern is a Gaussian function. There are no longer any rings, having a regularly decreasing amplitude, surrounding the central spot. One speaks of this as apodisation. One may thus reduce the light diffracted in the neighborhood of the central spot, in order not to mask less luminous neighboring objects.

3.7   Translation in Its Own Plane of the Aperture *T* Which Delimits the Wavefront

The aperture T in the screen E_1 is subjected to an arbitrary translation in its own plane (Fig. 3.13). Let $f(u, v)$ be the amplitude produced by the aperture $F(\eta, \zeta)$ in its original position. If C is an arbitrary point

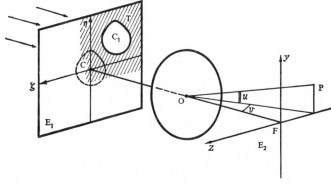

Fig. 3.13

in the aperture in this position, we assume that the translation transforms the point C into a point C_1 of coordinates η_1 and ζ_1. When one translates $F(\eta, \zeta)$ it is known (Appendix A, Sections A.2 and A.12) that its Fourier transform is multiplied by $\exp[jK(u\eta_1 + v\zeta_1)]$. The amplitude at P is then:

$$f_1(u, v) = e^{jK(u\eta_1 + v\zeta_1)}f(u, v) \tag{3.16}$$

One still has $I = f_1 f_1^*$ and only the phase has been changed. Since the detectors are not sensitive to phase, the diffraction pattern will show no change either in intensity or in position.

3.8   Diffraction at Infinity by Two Identical Slits

The screen E_1 is pierced by two identical slits of width ζ' and of height $\eta' \gg \zeta'$ (Fig. 3.14). They are symmetrically placed with respect to C and the distance between their centers is ζ_0. The slits are illuminated by a plane wave which is parallel to the plane of the screen E_1. Since the two slits are thin, practically no light is diffracted except along the axis Fz. In accordance with (3.16), the amplitude at P is:

$$\left[e^{jKv\zeta_0/2} + e^{-jKv\zeta_0/2}\right] \text{sinc}\left(\frac{\pi v\zeta'}{\lambda}\right) \tag{3.17}$$

from which the intensity is

$$I = I_0 \cos^2\left(\frac{\pi v\zeta_0}{\lambda}\right) \text{sinc}^2\left(\frac{\pi v\zeta'}{\lambda}\right) \tag{3.18}$$

One may arrive at this result in a different way, by noting that the pair of slits is represented by *the convolution of a single slit with two delta*

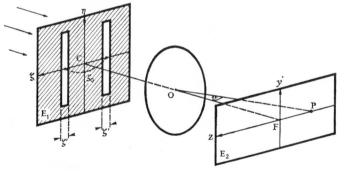

Fig. 3.14

functions. The Fourier transform is then given by the product of the transform of a slit centered at the origin with the transform of two delta functions, that is to say, with Young's fringes.

If one neglects diffraction, formula (3.18) shows that $I = I_0 \cos^2(\pi v \zeta_0/\lambda)$. The distribution of light along Fz is the same as that given by the interference of two coherent point sources separated by a distance ζ_0. Consequently, the diffraction by the slits modulates the interference pattern. Along the axis Fz the intensity distribution is represented by the curve of Fig. 3.15. In order to consider conditions identical to those of Section 1.8, it will be necessary to consider not two slits, but two very small identical apertures. If we assume that η' is equal to ζ', the intensity may be written

$$I = I_0 \cos^2\left(\frac{\pi v \zeta_0}{\lambda}\right) \operatorname{sinc}^2\left(\frac{\pi v \zeta'}{\lambda}\right) \operatorname{sinc}^2\left(\frac{\pi u \zeta'}{\lambda}\right) \qquad (3.19)$$

The interference phenomena are modulated by the diffraction pattern of a square. If ζ' is very small, the central part of the diffraction pattern is greatly spread out in two dimensions, and it is in this central part that one sees the interference fringes. These two very small apertures behave like the two coherent point sources M_1 and M_2 of Section 2.5. One can now understand why the interference phenomena are inseparable from the diffraction phenomena in this experiment. The wave divider here is very simply the screen E_1 itself pierced by the two small apertures.

As we remarked earlier, interference phenomena are not localized, but the phenomena of diffraction, themselves, are localized in the focal plane. So that if the slits are very thin, diffraction effects play a small role, and one readily sees interference fringes outside of the focal plane. But if the slits are not very thin, then as soon as one moves away

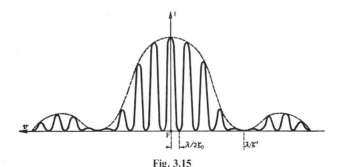

Fig. 3.15

from the focal plane, the diffraction pattern is modified, which also perturbs the interference fringes.

We may arrive at the preceding results by means of the remark made at the end of Section 3.3. It will suffice to calculate the auto-correlation function of the two slits and then to take its Fourier transform. One thus obtains directly the diffraction pattern in intensity.

3.9   Diffraction at Infinity by a Large Number of Identical Apertures Similarly Oriented and Irregularly Distributed

In accordance with (3.16), if N is the number of apertures, the amplitude at P is (Fig. 3.16):

$$f(u, v) \sum_i^N e^{jK(u\eta_n + v\zeta_n)} = f(u, v) \sum_i^N e^{jK\Delta_n} \tag{3.20}$$

where $f(u, v)$ represents the amplitude due to a single aperture. One may write (3.20) in the form:

$$f(u, v) \left[\sum_1^N \cos K\Delta_n + j \sum_1^N \sin K\Delta_n \right] \tag{3.21}$$

from which the intensity is:

$$I = |f(u, v)|^2 \left\{ \left[\sum_1^N \cos K\Delta_n \right]^2 + \left[\sum_1^N \sin K\Delta_n \right]^2 \right\} \tag{3.22}$$

or, continuing:

$$I = |f(u, v)|^2 \left\{ \sum_1^N (\cos^2 K\Delta_n + \sin^2 K\Delta_n) + 2 \sum_1^N \cos K(\Delta_n - \Delta_m) \right\} \tag{3.23}$$

Since there are a large number of apertures distributed at random, the last sum in the expression (3.23) takes on as many positive values as negative values, and it is to all intents and purposes equal to zero. Then one has:

$$I = N |f(u, v)|^2 \tag{3.24}$$

One has the same diffraction phenomena as with a single aperture, but N times brighter. As a matter of fact, observation of the phenomenon shows that it is necessary to study the problem in greater detail. The diffraction pattern produced by the group of apertures displays a granular structure which the preceding calculation does not

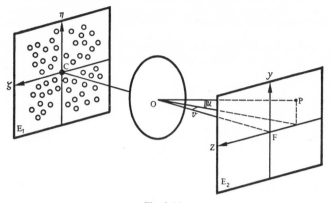

Fig. 3.16

explain. In the first place, at the center F, $\Delta_n = 0$ for all the apertures, the amplitude is equal to N and the intensity to N^2, if we set $f(0, 0) = 1$. Thus one observes at the center a small and very bright spot. Far from the axis, the phase difference $K\Delta_n$ take on various values between 0 and 2π. At another point, even one which is very close, all the phase differences change. Even if all the changes are small, they may produce at the end of the calculation significant variations in the amplitude and consequently in the intensity. These are the intensity fluctuations which give the phenomenon a granular appearance.

3.10   Complementary Screens. Babinet's Theorem

Consider two screens E_1 and E_1'; one screen, E_1, is formed of a large number of identical apertures, similarly oriented and distributed at random. The other, E_1', is formed of an equal number of small opaque screens having the same shape as the preceding apertures and occupying the same positions. The two screens E_1 and E_1' are complementary, the opaque parts of E_1' being just right for filling in exactly the open parts of E_1.

Let us first use the screen E_1': the amplitude $f_{E_1'}(u, v)$ at an arbitrary point P (Fig. 3.16) is equal to the amplitude $f(u, v)$ given by the objective O when it is unobstructed (no screen whatever) minus the amplitude $f_{E_1}(u, v)$ produced at P by the apertures which replace the small opaque screens

$$f_{E_1'}(u, v) = f(u, v) - f_{E_1}(u, v). \tag{3.25}$$

The diffraction pattern corresponding to $f_{E_1}(u, v)$ is much more spread out than the diffraction pattern $f(u, v)$ corresponding to the whole aperture of the objective O, since the apertures are very small. Therefore, if one makes observations far away from the center of the diffraction pattern $f(u, v)$ due to the entire aperture of the objective, $f(u, v)$ is negligible and one has:

$$f_{E_1'}(u, v) \cong -f_{E_1}(u, v) \qquad (3.26)$$

from which the intensity is:

$$|f_{E_1'}(u, v)|^2 \cong |f_{E_1}(u, v)|^2 \qquad (3.27)$$

At points distant from the diffraction phenomena produced by the entire surface area of the objective O, the diffraction patterns produced by two complementary screens are identical.

3.11   Elementary Properties of the Correspondence between the Aperture and the Diffraction Pattern

These properties may be deduced from the properties of the Fourier transformation in two dimensions. They appear implicitly in the preceding paragraphs and we shall summarize them.

(a)   *Rotation of the axes.* If one rotates the axes in their own plane around the axis COF (Fig. 3.8) (the same notation will be used in all the figures), the diffraction pattern is not changed, but rotates in its own plane through the same angle.

(b)   *Dilatation in a particular direction.* If the aperture is *dilated* in a certain direction, the diffraction pattern is *contracted* in the same direction.

(c)   *Homothety.* If one changes the aperture by a homothetic transformation, the diffraction pattern undergoes an inverse homothety. The smaller the aperture, the more the diffraction pattern becomes spread out.

(d)   *Translation.* We have seen (Section 3.7) that a translation of the aperture changes only the phase. The diffraction pattern is unchanged in intensity and in position.

3.12   Diffraction at a Finite Distance or Fresnel Diffraction

Let T be an aperture illuminated by a plane wave parallel to the plane E_1 of the aperture (Fig. 3.17). The dimensions of the aperture T, and the distance FP of the point P at which one calculates the amplitude,

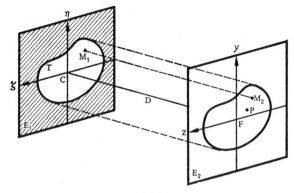

Fig. 3.17

are assumed to be small compared to D. In these conditions, the amplitude at P is given by (3.2):

$$f(y, z) = \frac{e^{-jK \cdot D}}{D} \iint\limits_{T} e^{-j(K/2D)[(y-\eta)^2 + (z-\zeta)^2]} \, d\eta \, d\zeta \qquad (3.28)$$

As we have said, studying the phenomenon of diffraction in the plane E_2 by means of the integral (3.28) is the same as studying the phenomena of diffraction at a finite distance or the Fresnel diffraction phenomena. One may represent the unobstructed part of the wavefront by a function $F(\eta, \zeta)$ which has a specified value in the interior of T and which is zero everywhere else. Omitting the constant factor outside the integral, the amplitude at P may be written:

$$f(y, z) = \int\limits_{-\infty}^{\infty}\!\!\!\int F(\eta, \zeta) e^{-j(K/2D)[(y-\eta)^2 + (z-\zeta)^2]} \, d\eta \, d\zeta \qquad (3.29)$$

The expression (3.29) shows that the amplitude $f(y, z)$ at P may be considered as a convolution of $F(\eta, \zeta)$ with $e^{-j(K/2D)[(y-\eta)^2 + (z-\zeta)^2]}$. We may write symbolically:

$$f(y, z) = F(y, z) \otimes e^{-jK(y^2 + z^2)/2D} \qquad (3.30)$$

Fresnel diffraction generally gives rise to more complicated calculations than do the phenomena at infinity (Fraunhofer).

3.13   Fresnel Zone Plate

The following example displays an application of formula (3.28). Let A_0 be a point source of monochromatic light which illuminates a photographic plate H (Fig. 3.18). Superpose on the spherical wave Σ emitted from A_0 a plane wave Σ_R. The two waves Σ and Σ_R are taken to be coherent. They thus originate from the same point source by means of an experimental arrangement which is not shown in Fig. 3.18,

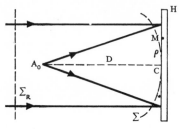

Fig. 3.18

in order not to complicate this diagram. The two waves Σ and Σ_R interfere and the intensity at an arbitrary point M located at a distance ρ from F is given by (2.8) which may be written here:

$$I = I_0 \cos^2 \frac{K\rho^2}{4D} \tag{3.31}$$

since the path difference at M between Σ and Σ_R is, to a good approximation, equal to $\rho^2/2D$. The interference fringes are annular rings centered on C (Fig. 3.19). The formula assumes that Σ and Σ_R are in phase at F, which does not in the least reduce the generality of the calculations which will follow. The energy W received by the photographic plate is equal to the product of the intensity by the exposure time τ:

$$W = \tau I_0 \cos^2 \frac{K\rho^2}{4D} = \frac{\tau I_0}{2}\left(1 + \cos \frac{K\rho^2}{2D}\right) \tag{3.32}$$

We shall see in Chapter 8 that if the variations in intensity are not too great, then, after development, the *amplitude* transmitted by the negative is proportional to W, the energy received. Formula (3.32) corresponds to interference fringes which have maximum contrast, since the minima are zero. In order that the variations in I and consequently in W shall not be too large, it will suffice that the amplitudes

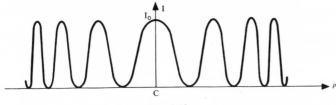

Fig. 3.19

of the waves Σ and Σ_R shall be different (Section 2.5). The fringes no longer have a maximum contrast, and one may write:

$$W = \frac{\tau I_0}{2}\left(m + \cos\frac{K\rho^2}{2D}\right) \tag{3.33}$$

where m is a constant greater than unity. The *amplitude t* transmitted by the photographic plate after development being proportional to W, one has:

$$t = a - bW \tag{3.34}$$

Since we are dealing with a negative, it is necessary that t decrease when W increase, and that is why we have placed the minus sign in front of the constant b. Setting:

$$t_0 = a - b\frac{\tau I_0}{2}m \qquad \text{and} \qquad \beta' = b\frac{\tau I_0}{2}$$

the amplitude transmitted may be written

$$t = t_0 - \beta'\cos\frac{K\rho^2}{2D} \tag{3.35}$$

The negative thus obtained constitutes a Fresnel zone plate. Illuminate the negative H located at E_1 with a bundle of parallel rays (Fig. 3.20).

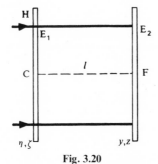

Fig. 3.20

We shall calculate the amplitude in a plane E_2 located at a distance l from E_1. In order to simplify the calculations, we shall calculate the amplitude on the axis of the annular rings at F. According to (3.29) one has here:

$$f(0, 0) = \int_0^{\rho_0} \int_0^{2\pi} F(\rho)e^{-j(K\rho^2/2l)}\rho \, d\rho \, d\theta \qquad (3.36)$$

where we have set $\rho^2 = \eta^2 + \zeta^2$ and where ρ_0 is the radius of the illuminated region of the negative H. In the expression (3.36) the function $F(\rho)$ represents the amplitude distribution on H which is given by (3.35). Integrating, one obtains to within a constant factor $\pi\rho_0^2 e^{j(K\rho_0^2/4l)}$:

$$f(0, 0) = t_0 \operatorname{sinc}\left(\frac{K\rho_0^2}{4l}\right) - \frac{\beta'}{2} e^{j(K\rho_0^2/4D)}\operatorname{sinc}\left[\frac{K}{4}\left(\frac{1}{D} + \frac{1}{l}\right)\rho_0^2\right]$$

$$- \frac{\beta}{2}e^{-j(K\rho_0^2/4D)}\operatorname{sinc}\left[\frac{K}{4}\left(\frac{1}{D} - \frac{1}{l}\right)\rho_0^2\right] \qquad (3.37)$$

The first term tends toward t_0 as l increases and it corresponds to the beam directly transmitted by H. The second and the third terms show maxima for $l = -D$ and $l = +D$ respectively (Fig. 3.21). One thus has two images of the source at a distance D from H: a real image and a virtual image. The negative we have constructed behaves somewhat like a lens, but with the difference that it gives rise to three images, one at infinity (the directly transmitted beam) and two images symmetrically placed with respect to the negative, one real and one virtual.

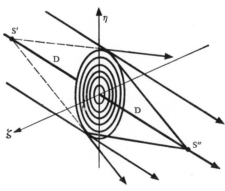

Fig. 3.21

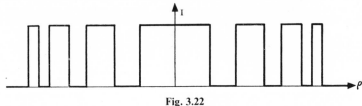

Fig. 3.22

Repeat the experiment with a grating having the profile shown in Fig. 3.22. One may represent it by the expression:

$$I = a + \frac{2}{\pi}[\sin \pi a \cos b\rho^2 + \tfrac{1}{2}\sin 2\pi a \cos 2b\rho^2 + \cdots] \quad (3.38)$$

Each term behaves like a circular grating analogous to that of Fig. 3.19, and gives two images of the source symmetrically placed with respect to the grating. When illuminated by a parallel bundle, this new circular grating thus gives rise to an infinite number of images of the source.

CHAPTER 4

Diffraction Gratings

4.1 Grating Composed of an Infinity of Parallel Slits

Let us consider a grating formed of an infinity of parallel equidistant slits of width ζ' (Fig. 4.1). The distance between centers of two consecutive slits (grating step) is equal to ζ_0. The grating is illuminated by a plane wave parallel to the plane E_1 of the slits, and one observes the diffraction phenomena in the focal plane of an objective O. It is known that in this case (Section 3.5) the diffracted light spreads out along the axis Fz and one treats the phenomena as having only a single

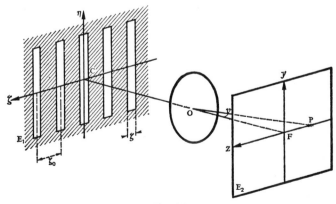

Fig. 4.1

parameter. Let Rect(ζ/ζ') be the rectangle function representing a slit. One may describe this situation by noting that the grating may be obtained by translating the solitary function Rect(ζ/ζ') through distances which are integer multiples of ζ_0. Since translation is equivalent to convolution with a delta function (appendix A, Section 14), then if $F(\zeta)$ represents the grating one has:

$$F(\zeta) = \text{Rect}\left(\frac{\zeta}{\zeta'}\right) \otimes \sum_{n=-\infty}^{+\infty} \delta(\zeta - \eta\zeta_0). \qquad (4.1)$$

or

$$F(\zeta) = \text{Rect}\left(\frac{\zeta}{\zeta'}\right) \otimes \text{comb}\left(\frac{\zeta}{\zeta_0}\right) \qquad (4.2)$$

The phenomenon of diffraction in amplitude is given by the Fourier transform of $F(\zeta)$. The transform of Rect(ζ/ζ') is (Section 3.5):

$$\text{F.T.}\left[\text{Rect}\left(\frac{\zeta}{\zeta'}\right)\right] = \text{sinc}\left(\frac{\pi v \zeta'}{\lambda}\right) \qquad (4.3)$$

The transform of the Dirac comb, comb(ζ/ζ_0), is another Dirac comb of step $v_0 = \lambda/\zeta_0$ from which:

$$f(v) = \text{F.T.}[F(\zeta)] = \text{sinc}\left(\frac{\pi v \zeta'}{\lambda}\right)\text{comb}\left(\frac{v}{v_0}\right) \qquad (4.4)$$

The interference pattern (Section 2.6) represented by comb(v/v_0) is modulated by the diffraction pattern of a slit, sinc($\pi v \zeta'/\lambda$). Figure 4.2. shows the intensity distribution $I = |f(v)|^2$ for $\zeta' \ll \zeta_0$ and for $\zeta_0 = 2\zeta'$.

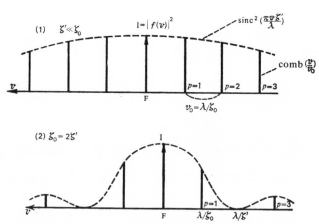

Fig. 4.2

When $\zeta' \ll \zeta_0$, the intensity of the spectra decreases slowly, and when $\zeta_0 = 2\zeta'$ all the spectra of even order disappear. The case $\zeta_0 = 2\zeta'$ corresponds to a grating in which the slits have a width equal to that of the opaque intervals which separate them. We may compare these results to those of Section 2.6.

If the source is no longer a luminous point at infinity, but instead is a thin slit parallel to the lines of the grating, there will be a superposition of the phenomena corresponding to each of the (incoherent) points of the source slit. The spectra have the form of bright lines parallel to the slits.

4.2 Grating Composed of a Finite Number of Slits

It will suffice to consider the infinite grating of the last section, and to multiply it by a rectangle function corresponding to the width L of the grating (Fig. 4.3). Figure 4.3 gives the intensity distribution $|F(\zeta)|^2$ in the plane of the grating. Formula (4.2) may be written here:

$$F(\zeta) = \left[\mathrm{Rect}\left(\frac{\zeta}{\zeta'}\right) \otimes \mathrm{comb}\left(\frac{\zeta}{\zeta_0}\right) \right] \mathrm{Rect}\left(\frac{\zeta}{L}\right) \qquad (4.5)$$

The diffraction pattern for the amplitude, according to (4.4), is:

$$f(v) = \mathrm{F.T.}[F(\zeta)] = \left[\mathrm{sinc}\left(\frac{\pi v \zeta'}{\lambda}\right) \mathrm{comb}\left(\frac{v}{v_0}\right) \right] \otimes \mathrm{sinc}\left(\frac{\pi v L}{\lambda}\right) \quad (4.6)$$

which one may write:

$$\left[\sum_{n=-\infty}^{+\infty} \mathrm{sinc}\left(\frac{\pi v \zeta'}{\lambda}\right) \cdot \delta(v - nv_0) \right] \otimes \left(\frac{\pi v L}{\lambda}\right) \qquad (4.7)$$

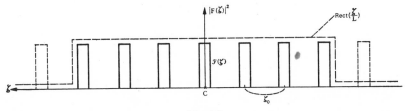

Fig. 4.3

and since convolution is distributive with respect to addition, one has further:

$$\sum_{n=-\infty}^{+\infty} \left[\operatorname{sinc}\left(\frac{\pi v \zeta'}{\lambda}\right) \cdot \delta(v - nv_0) \otimes \operatorname{sinc}\left(\frac{\pi v L}{\lambda}\right) \right] \qquad (4.8)$$

In the case of a grating, $L \gg \zeta_0$ and consequently the function $\operatorname{sinc}(\pi v L/\lambda)$, as a practical matter, extends only over an interval which is very small compared to $v_0 = \lambda/\zeta_0$. The function $\operatorname{sinc}(\pi v L/\lambda)$ is simply translated by integer multiples of $v_0 = \lambda/\zeta_0$, its ordinates being multiplied by a constant factor which depends only on the translation. For a translation equal to nv_0, this constant factor equals $\operatorname{sinc}(\pi v \zeta'/\lambda) \cdot \delta(v - nv_0)$. Figure 4.4 represents the intensity distribution

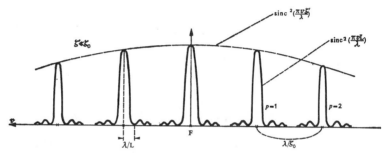

Fig. 4.4

of the spectra. It is the finite width L of the grating which gives rise to the finite width of the spectra. Each spectrum has the structure of the diffraction pattern of a slit of width L equal to the width of the grating. We have assumed $\zeta' \ll \zeta_0$ in Fig. 4.4. The half-width of the diffraction pattern constituting each spectrum is λ/L. Suppose now that the grating is illuminated by a different wavelength λ'. For the wavelength λ the position of the pth spectrum is $p\lambda/\zeta_0$, according to Fig. 4.4 (Section 3.6), and it is $p\lambda'/\zeta_0$ for the wavelength λ'. The angular spacing between two "lines" of wavelengths λ and λ' is, for the pth order spectrum:

$$p\frac{\lambda}{\zeta_0} - p\frac{\lambda'}{\zeta_0} = \frac{p}{\zeta_0}(\lambda - \lambda') \qquad (4.8a)$$

This formula permits one to determine the minimum difference $d\lambda = \lambda - \lambda'$ of the wavelengths of the two lines which the grating is capable of separating. Recall the following classic test: two lines are at their limit of resolution if the maximum of one of the lines corresponds

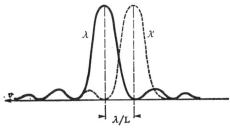

Fig. 4.5

to the first zero of the other line (Fig. 4.5). When the two lines are related thus, the distance of their centers is λ/L. In accordance with (4.8a) one therefore has:

$$\frac{p}{\zeta_0}(\lambda - \lambda') = \frac{p\,d\lambda}{\zeta_0} \geq \frac{\lambda}{L} \tag{4.9}$$

from which we find:

$$d\lambda \geq \frac{\lambda\zeta_0}{p\zeta} \tag{4.10}$$

The ratio $R = \lambda/d\lambda$ is called *the resolving power* of the grating. Using (2.13), we find:

$$R = \frac{vL}{\lambda} \tag{4.11}$$

which depends only on the width L of the grating and the angle v at which it is used. In practice, one often uses angles v which are not small, and the formula $v = p\lambda/\zeta_0$ is replaced by $\sin v = p\lambda/\zeta_0$. In the same way, one can take account of a possible inclination of the incident wave.

4.3   Blazed Grating

For simplicity, we shall suppose to begin with that the grating is unbounded, as in Section 4.1. The new grating is formed of transparent slits within which the phase varies linearly from zero to φ_0. All the "slits" are joined together in sequence, and there no longer are any opaque intervals. One then has a transparent *phase* grating whose

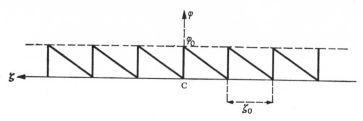

Fig. 4.6

phase variations have the profile shown in Fig. 4.6. Such a profile may
be obtained in a grating composed of small parallel bands of transparent material having a prismatic cross-section. This grating may be
considered as the convolution of a Dirac comb of step ζ_0, with the
complex function $\mathcal{F}(\zeta)$ representing the variation of phase in each strip.
We shall make use of formula (4.2). We shall determine the function
$\mathcal{F}(\zeta)$ and calculate its Fourier transform. Figure 4.7 represents the
prismatic cross-section of a single one of the strips of the grating. Let α
be the angle of the prisms and n their index of refraction.

At the distance ζ from the point O the path difference is $(n-1)\overline{MM'}$
$= (n-1)\alpha\zeta$, and the phase difference is $K(n-1)\alpha\zeta$. The condition
imposed earlier gives:

$$K(n-1)\alpha\zeta_0 = \varphi_0 \tag{4.12}$$

One then has:

$$\mathcal{F}(\zeta) = e^{jK(n-1)\alpha\zeta} \tag{4.13}$$

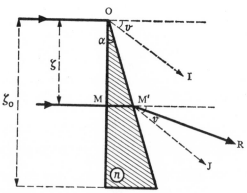

OI and $M'J$ are rays *diffracted* in the direction v
$M'R$ is the *refracted* ray corresponding to the incident ray MM'.

Fig. 4.7

As a result of the phase variations, the strips diffract light and the Fourier transform of $\mathcal{F}(\zeta)$ is:

$$\int_0^{\zeta_0} e^{jK(n-1)\alpha\zeta} e^{jKv\zeta}\, d\zeta = e^{jK[v+(n-1)\alpha]\zeta_0/2}\, \text{sinc}\left\{K[v+(n-1)\alpha]\frac{\zeta_0}{2}\right\}$$

(4.14)

One may represent the grating by an expression analogous to (4.2):

$$F(\zeta) = e^{jK(n-1)\alpha\zeta} \otimes \text{comb}(\zeta/\zeta_0)$$

(4.15)

In order to determine the amplitude in the focal plane of the objective O, it is necessary to form the Fourier transform of the two members of (4.15).

Leaving to one side the exponential of (4.14), one has:

$$f(v) = \text{sinc}\left\{K[v+(n-1)\alpha]\frac{\zeta_0}{2}\right\} \cdot \text{comb}\left(\frac{v}{v_0}\right)$$

(4.16)

The only way in which this differs from Section 4.1 is in the shifting of the diffraction pattern due to a single slit. This pattern is no longer centered at F. We shall make use of this result in the following way: let us give φ_0 a value such that the maximum of the diffraction pattern of a slit coincides with, for example, the first order spectrum.

From (4.16) the abscissa of the maximum is:

$$v_m = -(n-1)\alpha$$

(4.17)

and the first zero corresponds to:

$$v = \frac{\lambda}{\zeta_0} - (n-1)\alpha$$

(4.18)

If one sets $(n-1)\alpha = \lambda/\zeta_0$ one sees that all the spectra coincide with the minima of the diffraction pattern of a slit, except for the spectrum of order 1. One has the arrangement shown in Fig. 4.8. The grating gives only a single spectrum, the spectrum of first order. One says that the spectrum is blazed in the first order. This is a result which is important in spectroscopy, since the energy is concentrated in a single spectrum, instead of being distributed over a number of spectra, whether large or small. From (4.17) one sees that the position of the spectrum corresponds to the direction of rays refracted by a prism of angle α and index n. In this paragraph, we have not taken account of the finite width L of the grating, which enters in the same fashion as before

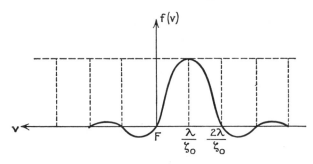

Fig. 4.8

(Section 4.2). It suffices to multiply (4.15) by a rectangle function, and then to carry out the same calculation as in Section 4.2.

In contrast to what one might expect, the practical construction of such a grating is not impossible. Blazed reflection gratings are currently manufactured. High resolution photographic plates permit one to make transparent gratings having the properties which have just been described.

4.4   Construction of Gratings. Ghosts

Gratings are made by using a diamond to rule equidistant grooves on a plate of glass. In spectroscopy, one often uses reflection gratings. In order to obtain these gratings, one rules on a flat metallic surface regular grooves which behave like opaque bands. One may rule the grating on a thin opaque metallic layer which has been deposited by evaporation in a vacuum on to a plane sheet of glass. In reflection gratings, the glass support is not traversed by light and any inhomogeneities in the glass have no effect, which is an advantage as compared with transmission gratings. Moreover, reflection gratings can be used in the ultraviolet or the infrared when a glass support could not be used in these regions.

A replica of an original grating may be obtained by allowing a solution of collodion in ether to flow over the grating. After evaporation of the ether, one may remove from the grating a transparent film which reproduces the grooves of the original grating. A transparent replica may be transformed into a replica usable for reflection by vacuum deposition of a thin metallic layer. Gratings employed in spectroscopy

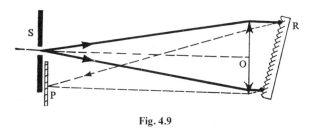

Fig. 4.9

generally are ruled with 500 lines per millimeter, and the grooved area varies from approximately $10 \times 10 \text{ cm}^2$ to $25 \times 25 \text{ cm}^2$.

One may employ gratings in different experimental arrangements. The Littrow type mounting (Fig. 4.9) is one of those most often employed. The slit S is at the focal point of the objective O. The reflection grating R, placed behind O, is oriented in such a way that the spectrum to be studied falls on the photographic plate P located near S, which is at the focal point of O. One thus has an autocollimating spectrograph. In making gratings, the number of grooves corresponding to a single turn of the generating screw is very large. If the generating screw is slightly off-center, the grooves will no longer be equidistant and the defect will be repeated in identical fashion for each turn of the screw. Homologous points of the reflecting grooves will no longer be equidistant. They are displaced with respect to their theoretical position, which will affect the position of the diffracted wavelets and consequently the form of the envelope of the wavelets, which will no longer be plane. The wavefront diffracted in an arbitrary direction will manifest a periodic deformation. If the deformation is sinusoidal, one may easily show by means of the Fourier transformation that there will appear on each side of a principal line two lines of the same wavelength as the principal line. These two lines are called "ghosts". If the deformation of the wavefront is an arbitrary periodic deformation, one may consider the deformation as a sum of sinusoidal deformations. For each component sinusoidal deformation, one has two ghosts, and consequently the principal line is surrounded on each side by a whole series of parasitic lines (the ghosts) each corresponding to the same wavelength. These ghosts are, of course, detrimental to the quality of the grating.

It is pertinent to point out that the use of the laser and photographic plates, particularly thermoplastics, allows us nowadays to produce excellent gratings by interference of two plane waves.

4.5 Sinusoidal Phase Grating

Gratings characterized by variations of absorption are called *amplitude gratings*. A phase grating is formed by periodic variations of phase without variation of absorption. The grating is uniformly transparent (or reflecting). Let us consider the case of a transparent sinusoidal grating. The phase variations are given by an expression of the form $e^{ja\sin(2\pi\zeta/\zeta_0)}$, where ζ_0 is the grating step and a is a constant which fixes the maximum value of the phase variation. The profile of the phases of this grating is given in Fig. 4.10. If L is the width of the grating, this curve may be represented by the function:

$$F(\zeta) = e^{ja\sin(2\pi\zeta/\zeta_0)}\,\text{Rect}(\zeta/L) \tag{4.19}$$

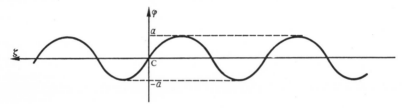

Fig. 4.10

In the focal plane of the objective used for observation (the experimental arrangement may be that of Fig. 4.1) the amplitude on the axis Fz is:

$$f(v) = \text{sinc}\left(\frac{\pi v L}{\lambda}\right) \otimes \sum_{n\infty-\infty}^{+\infty} J_n(a) \cdot \zeta\left(v - \frac{n\lambda}{\zeta_0}\right) \tag{4.20}$$

$J_n(a)$ being the Bessel function of order n and argument a, from which

$$f(v) = \sum_{n=-\infty}^{+\infty} J_n(a) \,\text{sinc}\left[\frac{\pi L}{\lambda}\left(v - \frac{n\lambda}{\zeta_0}\right)\right] \tag{4.21}$$

Since $\zeta_0 \ll L$, the terms of the series practically do not overlap and one may write:

$$I = \sum_{n=-\infty}^{+\infty} J_n^2(a) \,\text{sinc}^2\left[\frac{\pi L}{\lambda}\left(v - \frac{n\lambda}{\zeta_0}\right)\right] \tag{4.22}$$

One finds once again the spectra of Fig. 4.4 corresponding to formula (4.7), but the modulation here is controlled by the function $J_n^2(a)$.

The disposition and the width of the spectra are unchanged, and only the intensities of the spectra are changed. The intensity of the central line located at F ($n = 0$) is proportional to $J_0^2(a)$. If a is any root whatever of the Bessel function of order zero, there will be no central line.

4.6 Two-Dimensional Gratings

We shall consider an amplitude grating, that is to say, a grating characterized only by variations in absorption. It is composed of small rectangular apertures of sides η' and ζ' (Fig. 4.11). Parallel to $c\eta$ the

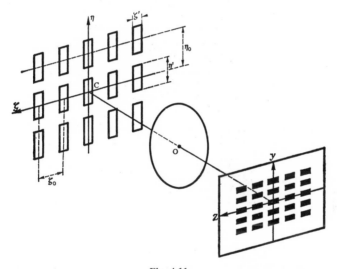

Fig. 4.11

grating step is η_0 and parallel to $c\zeta$ it is ζ_0. This grating may be represented by the function:

$$F(\eta, \zeta) = \left[\mathrm{comb}\left(\frac{\eta}{\eta_0}\right)\mathrm{comb}\left(\frac{\zeta}{\zeta_0}\right)\right] \otimes \left[\mathrm{Rect}\left(\frac{\eta}{\eta'}\right)\mathrm{Rect}\left(\frac{\zeta}{\zeta'}\right)\right] \quad (4.23)$$

from which the amplitude diffraction pattern is given by:

$$f(u, v) = \left[\mathrm{comb}\left(\frac{u}{u_0}\right)\mathrm{comb}\left(\frac{v}{v_0}\right)\right]\left[\mathrm{sinc}\left(\frac{nu\eta'}{\lambda}\right)\mathrm{sinc}\left(\frac{nv\zeta'}{\lambda}\right)\right] \quad (4.24)$$

This formula may readily be interpreted in terms of the previous results. The diffraction pattern consists of a checkerboard arrangement of spectra. The smaller η' and ζ' are, the more slowly do the intensities decrease.

4.7   Diffraction by a Crystal

It is well known that in a crystal the periodicity of the atoms is characterized by a crystal lattice. For simplicity, we consider a crystal in which all the atoms are identical. Figure 4.12 corresponds to a model in two dimensions. The elementary cell of the lattice has the form of a parallelogram in Fig. 4.12. The group of atoms located within the

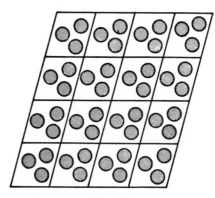

Fig. 4.12

elementary cell is the motif. It is the repetition of this motif in three-dimensional space which generates the crystal. The crystalline lattice is much too closely spaced to diffract light, but is very well suited for observing the diffraction of x rays. Given the periodic structure of the crystalline medium, one may regard a crystal as the convolution of the motif by delta functions located at the lattice points. The motif itself is represented by the convolution of an atom by a series of delta functions corresponding to the positions of the various atoms in the elementary cell. Consequently, the diffraction pattern of a crystal is the following product: the transform of an atom times the transform of the delta functions corresponding to the positions of the atoms within the elementary cell, times the transform of the set of delta functions located at the lattice points.

CHAPTER 5

Partial Coherence

5.1 Quasi-Monochromatic Vibrations

According to Fourier's theorem, a function which is integrable and which is everywhere finite may be represented as the sum of a continuous infinity of sinusoidal components of different frequencies v. If one agrees to allow the frequency to vary from $-\infty$ to $+\infty$, the vibration $U(t)$ due to a solitary train of waves (a pulse) is represented in complex notation by the Fourier integral:

$$U(t) = \int_{-\infty}^{\infty} v(v)e^{j2\pi vt} \, dv \qquad (5.1)$$

Here $v(v)$ specifies the amplitude and the phase of each monochromatic component: it is the spectrum of the vibration. The energy distribution of these components is given by $|v(v)|^2$. The Fourier transform may be written:

$$v(v) = \int_{-\infty}^{+\infty} U(t)e^{-j2\pi vt} \, dv \qquad (5.2)$$

This expression permits one to calculate $v(v)$, the spectrum of $U(t)$. If $v(v)$ is nonzero only for values of v near a mean value v_0 it is convenient to write $U(t)$ in the following form:

$$U(t) = e^{j2\pi v_0 t} \int_{-\infty}^{\infty} v(v)e^{j2\pi(v - v_0)t} \, dv \qquad (5.3)$$

59

and setting:

$$a(t) = \int_{-\infty}^{\infty} v(v)e^{j2\pi(v-v_0)t} \, dv \qquad (5.4)$$

one has

$$U(t) = a(t)e^{j2\pi v_0 t} \qquad (5.5)$$

where $a(t)$ varies relatively slowly as a function of time since $v - v_0$ always remains small. More precisely, $a(t)$ varies slowly compared with $\cos 2\pi v_0 t$ or $\sin 2\pi v_0 t$. The variations of $a(t)$ are slow compared to the period of vibration $T = 1/v_0$, but it is important to point out that these are still extraordinarily rapid compared to the usual methods of observation. The vibration $U(t)$, defined in this way, is a quasi-monochromatic vibration. It differs from a monochromatic vibration $e^{j2\pi v_0 t}$ by the presence of the factor $a(t)$. Let us sketch in the complex plane the complex amplitude $a(t)$ of the vibration for different types of vibrations. In the case of a monochromatic vibration, $a(t)$ is constant and is represented by a point such as M (Fig. 5.1). For a damped sinusoidal vibration one has:

$$a(t) = e^{-t/\tau}e^{j\theta} \qquad (5.6)$$

where τ is the damping constant (the lifetime in quantum theory) and θ is the phase of the vibration. The point M representing $a(t)$ moves in toward the origin along the straight line OM without undergoing any rotation, since θ is constant for the wavetrain (Fig. 5.2). If one considers the Doppler effect, there is an apparent variation of the frequency Δv due to the speed V of the atom with respect to the observer:

$$\frac{\Delta v}{v} = \frac{V}{c} \qquad (5.7)$$

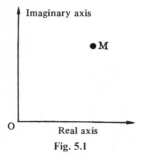

Fig. 5.1

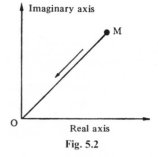

Fig. 5.2

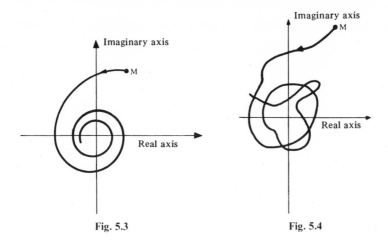

Fig. 5.3   Fig. 5.4

the atom appears to emit a frequency $v_0 + \Delta v$ and for a damped vibration one has:

$$a(t) = e^{-t/\tau} e^{j\theta} e^{j2\pi \Delta v t} \tag{5.8}$$

The phase varies with the time, and in the complex plane one has the curve of Fig. 5.3. Finally, in the general case, the amplitude $a(t)$ varies in a complicated fashion, and one can have a curve similar to that of Fig. 5.4.

5.2 Relation between the Length of a Wavetrain (Coherence Length) and the Line Width of the Radiation Emitted

According to (5.2) the spectrum $v(v)$ of the vibration $U(t)$ is given by the Fourier transform of $U(t)$. Take, for example, a sinusoidal vibration having a finite duration τ (coherence time). One has

$$U(t) = e^{j2\pi v_0 t}, \qquad |t| \leq \frac{\tau}{2}$$
$$\tag{5.9}$$
$$U(t) = 0, \qquad |t| > \frac{\tau}{2}$$

The real part of $U(t)$ is shown in Fig. 5.5. From (5.2) one has

$$v(v) = \int_{-\tau/2}^{+\tau/2} e^{-j2\pi(v - v_0)t} \, dt = \tau \, \frac{\sin \pi(v - v_0)\tau}{\pi(v - v_0)\tau} \tag{5.10}$$

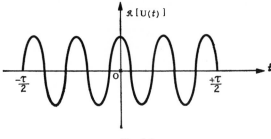

Fig. 5.5

The profile of the energy spectrum $|v(v)|^2$ is given by the curve of Fig. 5.6.

If the light source emits, at random, wavetrains which are altogether identical to the preceding one, one may assume that its spectrum is the same as that of the light emitted by the source.

The interval $2\,\Delta v$ of the frequencies between points A and B (Fig. 5.6) that are located symmetrically with respect to v_0 may serve to

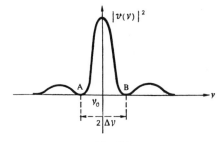

Fig. 5.6

characterize the spectral width of the light emitted. According to (5.10) one has:

$$\Delta v = \frac{1}{\tau} \tag{5.11}$$

The spectral width is of the order of the reciprocal of the duration of the wavetrain, that is to say, of the coherence time. To the frequency interval Δv there corresponds a wavelength interval $\Delta \lambda$ which may be calculated from the relation $\lambda = c/v$. To within a sign one has:

$$\Delta \lambda = c \frac{\Delta v}{v^2} = \Delta v \frac{\lambda^2}{c} \tag{5.12}$$

from which the length of the wavetrain (coherence length) is:

$$l = c\tau = \frac{c}{\Delta v} = \frac{\lambda^2}{\Delta\lambda} \tag{5.13}$$

For the green line of a mercury vapor lamp at moderate pressure, $\lambda = 546.1$ nm and the line width is $\Delta\lambda = 30$ nm; then one has $l = 10$ μm. The longer the wavetrain and the narrower the spectrum, the more light approximates to a monochromatic radiation. In the limit, an infinite wavetrain corresponds to a monochromatic radiation.

5.3   Remark

One may also obtain the preceding results by starting with the integral (5.4) and writing:

$$v(v) = \int_{-\infty}^{\infty} a(t)e^{j2\pi v_0 t}e^{-j2\pi v t}\, dt \tag{5.14}$$

The spectrum $v(v)$ is given by the Fourier transform of $U(t) = a(t)e^{j2\pi v_0 t}$, which is equal to the convolution of the transforms of $a(t)$ and $e^{j2\pi v_0 t}$. It will thus suffice to calculate the transform of $a(t)$, and to subject it to a translation v_0. In the calculation of the preceding Section 5.2, it was not necessary to introduce the phase θ since we were considering a wavetrain. The amplitude $a(t)$ is thus real, and it represents the "envelope" of the vibration. This is a rectangle function corresponding to a duration τ and one has:

$$a(t) = \text{Rect}\left(\frac{t}{\tau}\right) \tag{5.15}$$

On subjecting the transform of $a(t)$ to a translation v_0 one again arrives at the result given by (5.10). Finally, take note of the fact that one can calculate $|v(v)|^2$ directly starting with the autocorrelation function of $a(t)$ (Section 3.3).

5.4   Receiver Illuminated by a Thermal Source

Consider now the case of a receiver having a resolution time which is large compared to the coherence time τ. This is generally the case for thermal sources. During the time needed for an observation, the

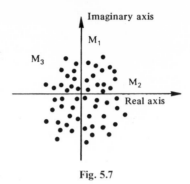

Fig. 5.7

receiver is illuminated by a considerable number of wavetrains. If the wavetrains are finite sinusoids such as those we have considered earlier (Section 5.2), then for each wavetrain, $a(t)$ may be represented in the complex plane (Fig. 5.7) by a point during the lifetime of the wavetrain considered. The receiver will absorb a very large number of wavetrains during the period of time needed for an observation. One may represent these by points such as M_1, M_2, M_3, ... distributed at random. The phases θ corresponding to different wavetrains have no relation whatever one to the other. The receiver can respond only to the average of the effects produced by these vibrations. However, the receivers, whether the eye, a photoelectric cell, a photographic plate, etc., are quadratic receptors which are responsive to energy. To the instantaneous complex amplitude $a(t)$ there corresponds the energy $|a(t)|^2$ which is the square of the modulus of $a(t)$. It is this quantity which determines the intensity of the phenomenon. For the receiver, the intensity of the phenomenon is thus characterized by the average value:

$$\overline{|a(t)|^2} = \overline{|U(t)|^2} = \overline{a(t)a^*(t)} = \frac{1}{2T} \int_{-T}^{+T} |a(t)|^2 \, dt \qquad (5.16)$$

where $2T$ is the time needed to carry out an observation.

5.5 Spatial Coherence in the Young Experiment

We shall assume that the experiment is carried out in a vacuum where all frequencies propagate with the same speed. The source S which illuminates the two holes T_1 and T_2 (Fig. 5.8) is an extended incoherent source, that is to say, a source each of whose elements are

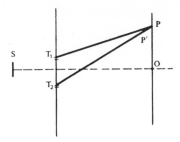

Fig. 5.8

considered as incoherent point sources. The source S emits light that is quasi-monochromatic.

Let us consider the vibrations emitted at the same instant by T_1 and T_2. At the instant t, the vibration emitted by T_1 and arriving at P is represented by the complex amplitude $a_1(t)$. At the instant t, the vibration emitted by T_2 will not arrive at P but for example at P' and its amplitude is $a_2(t)$. What happens at P at the time t is then the same as what happens at P' at time $t - \theta$, and it is represented by the expression $a_2(t - \theta)e^{-j2\pi v_0\theta}$, where θ is the time taken for the vibration to travel from P' to P. In what follows, we shall assume that the path difference $\overline{T_2P} - \overline{T_1P} = \overline{PP'} = c\theta$ is always much smaller than the length of the wavetrains. As in other respects $a_1(t)$ and $a_2(t)$ vary slowly compared with the period of vibration, one may set $a_2(t - \theta) = a_2(t)$. The complex amplitude at P may then be written:

$$a(t) = a_1(t) + a_2(t)e^{-j2\pi v_0\theta} \tag{5.17}$$

From (5.16) and to within a constant factor, the intensity at P is, after some simplification of the notation:

$$I = \overline{a(t)a^*(t)} = \overline{(a_1 + a_2 e^{-j2\pi v_0\theta})(a_1^* + a_2^* e^{j2\pi v_0\theta})} \tag{5.18}$$

where $2\pi v_0\theta = 2\pi c\theta/\lambda_0$ represents the phase difference at P between the vibrations arriving at this point from T_1 and T_2. Applying the average value symbol only to those quantities which vary as functions of the time, one has:

$$I = \overline{a_1 a_1^*} + \overline{a_2 a_2^*} + \overline{a_1^* a_2}\,e^{-j\varphi} + \overline{a_1 a_2^*}\,e^{j\varphi} \tag{5.19}$$

where $\overline{a_1a_1^*}$ and $\overline{a_2a_2^*}$ represent the intensities I_1 and I_2 produced by T_1 and T_2 acting separately. One may write:

$$I = I_1 + I_2 + 2\,\text{Re}[\overline{a_1a_2^*}\,e^{j\varphi}] \tag{5.20}$$

where Re signifies the real part of the expression in brackets. If we introduce a normalizing factor and set:

$$\gamma_{12} = \frac{\overline{a_1(t)a_2^*(t)}}{\sqrt{I_1}\sqrt{I_2}} \tag{5.21}$$

then γ_{12} is the *complex degree of coherence* of the vibrations emitted by T_1 and T_2. If α is the argument of γ_{12} one has

$$\gamma_{12} = |\gamma_{12}|e^{j\alpha} \tag{5.22}$$

from which:

$$I = I_1 + I_2 + 2\sqrt{I_1}\sqrt{I_2}|\gamma_{12}|\cos(\alpha + \varphi) \tag{5.23}$$

$|\gamma_{12}|$ is *the degree of coherence* of the vibrations emitted by T_1 and T_2. From the Schwartz inequality $|\gamma_{12}| \le 1$ and there are three cases to consider:

1. $|\gamma_{12}| = 1$. Formula (5.23) becomes identical with the classical Fresnel formula (2.10). One says that the sources are *coherent*.
2. $|\gamma_{12}| = 0$. Formula (5.23) gives $I = I_1 + I_2$. There no longer are any interference phenomena at P. One says that the sources T_1 and T_2 are *incoherent*.
3. $0 < |\gamma_{12}| < 1$. The vibrations emitted by T_1 and T_2 are *partially coherent* and $|\gamma_{12}|$ measures their degree of coherence.

The fringe contrast may easily be calculated using formula (5.23). Making use of the expression (2.3) one finds, if $I_1 = I_2$:

$$\gamma = |\gamma_{12}| \tag{5.24}$$

The fringe contrast is given by the degree of coherence of T_1 and T_2, a degree of coherence which depends on the extension of the source S, and which we are going to calculate.

5.6   Degree of Coherence of Two Points T_1 and T_2 Illuminated by the Extended Source S

The source S is a plane incoherent source which emits quasi-monochromatic light (Fig. 5.9). We shall assume that T_1 and T_2 are in a plane parallel to the plane of the source. The distance T_1T_2 and

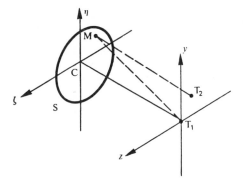

Fig. 5.9

the dimensions of the source are assumed to be small compared to the distance separating the source and the plane containing T_1 and T_2. For simplicity's sake the point T_1 is taken to coincide with the origin of coordinates y, z. If $a(t)$ is the amplitude at T_1 due to M, the amplitude at T_2 may be written $a(t)e^{-j2\pi v_0\theta}$, where $\theta = (\overline{MT_2} - \overline{MT_1})/c$ because we are still assuming that the path difference $MT_2 - MT_1$ is much smaller than the length of the wavetrain. The total amplitudes at T_1 and T_2 due to all the points of the source S will be:

$$a_1(t) = \Sigma a_i(t), \qquad a_2(t) = \sum a_j(t)e^{-j2\pi v_0\theta_j} \qquad (5.25)$$

from which:

$$\overline{a_1 a_2^*} = \overline{\Sigma a_i \Sigma a_j^* e^{j2\pi v_0\theta_j}} = \overline{\Sigma a_i a_i^*}\, e^{j2\pi v_0\theta_i} \qquad (5.26)$$

In fact, two points of the source are incoherent, and according to the preceding paragraph, their complex degree of coherence, which is proportional to $\overline{a_i a_j^*}$, is zero. Consequently, averages such as $\overline{a_i a_j^*}$, for which i is different from j, are zero. Besides, it is inconvenient to show the average symbol on the exponential, which does not depend on the time but only on the path difference $c\theta_i$. But $\overline{a_i a_i^*}$ represents the luminous intensity $J(\eta, \zeta)$ at the point M of the source, and since the source is divided into a large number of incoherent elements $d\eta\, d\zeta$, one may replace the sum by an integral:

$$\overline{a_1 a_2^*} = \iint_S J(\eta, \zeta)e^{j2\pi v_0\theta}\, d\eta\, d\zeta \qquad (5.27)$$

Now, according to Section 3.3, the phase difference $2\pi v_0 \theta$ here is equal to $K(\overline{MT_2} - \overline{MT_1})$, which is also given by the expression $K(u\eta + v\zeta)$. Thus one has on normalizing:

$$\gamma_{12} = \frac{\iint_S J(\eta, \zeta)e^{jK(u\eta + v\zeta)} \, d\eta \, d\zeta}{\iint_S J(\eta, \zeta) \, d\eta \, d\zeta} \tag{5.28}$$

The complex degree of coherence of two points T_1 and T_2 illuminated by the source S is given by the normalized Fourier transform of the intensity distribution $J(\eta, \zeta)$ over the source. The modulus of the expression (5.28) gives the degree of coherence of T_1 and T_2.

According to (5.28), the degree of coherence makes use of the amplitudes $a_1(t)$ and $a_2(t)$ calculated at T_1 and T_2. On the other hand, in the expression (5.21) the degree of coherence has been based on the values of the amplitudes $a_1(t)$ and $a_2(t)$ evaluated at the instant when they arrive at P (Fig. 5.8). This has no great importance, because this amounts to changing the origin of time and to neglecting the real decrease of amplitude with distance, a decrease which is practically the same for the two distances $T_1 P$ and $T_2 P$ in Fig. 5.8.

5.7 Contrast of the Interference Fringes in Young's Experiment

This calculation may be carried out immediately, starting from the expressions (5.24) and (5.28). Consider a uniform circular source, to which corresponds $J(\eta, \zeta) = 1$. From (3.12) and (3.13) one has:

$$\gamma_{12} = \frac{2J_1(Z)}{Z}, \qquad Z = \frac{2\pi}{\lambda}\alpha\overline{T_1 T_2} \tag{5.29}$$

α being the angle subtended by the radius of the source as seen from T_1. The fringe contrast is thus given by:

$$|\gamma_{12}| = \left|\frac{2J_1(z)}{z}\right| \tag{5.30}$$

This is zero if $J_1(Z) = 0$, that is to say if $Z = 3.83$, which gives:

$$\overline{T_1 T_2} = \frac{1.22\lambda}{2\alpha} \tag{5.31}$$

5.8   Michelson's Method for Measuring the Apparent Diameter of a Star

The light originating in the star is reflected by the four mirrors $T_1 T_2 M_1 M_2$ (Fig. 5.10), and then passes through the two slits F_1 and F_2 placed in front of the objective O. In the original experimental arrangement, the objective O was in fact the mirror of a telescope. In the focal plane of the objective O, one observes Young's fringes, whose spacing is determined by the distance $F_1 F_2$. One may replace by two points the two mirrors T_1 and T_2; if they are illuminated in an incoherent fashion by the star, the bundles which they reflect will not be able to interfere in the focal plane of the objective O. The fringe contrast depends on the degree of coherence of T_1 and T_2, as illuminated by the star. Formula (5.31) shows that α should be small, and the distance $T_1 T_2$ ought to be large, in order that the fringes shall disappear. The separation of the two mirrors is changed, and one observes the moment when the fringes contrast becomes zero. Formula (5.31) gives the apparent diameter 2α of the star.

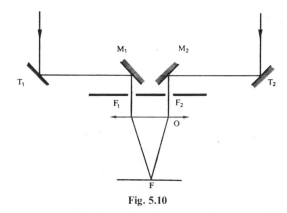

Fig. 5.10

5.9   Measurement of the Apparent Diameter of a Star by the Method of Stationary Waves

The light originating in the star is reflected by the two mirrors T_1 and T_2 (Fig. 5.11). The waves propagating in opposite directions give rise to a system of stationary waves, which are detected by the receiver R. One may once again replace the mirrors T_1 and T_2 by two points; if

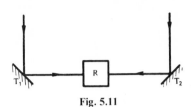

Fig. 5.11

they are incoherently illuminated by the star, the stationary waves will not be visible. The visibility of the fringes produced by the stationary waves depends on the degree of coherence of T_1 and T_2, as illuminated by the star. One increases the distance $T_1 T_2$ until the moment when there are no longer any stationary waves. Formula (5.31) gives the apparent diameter 2α of the star.

CHAPTER 6

Interference Phenomena
in Polarized Light

6.1 Luminous Vibration in an Isotropic Dielectric Medium. Natural Light

In all the preceding work, it was not necessary to consider the direction of the vibration, which could thus be considered as a scalar quantity. When one studies the phenomena of interference in polarized light, the situation is no longer the same, and it is necessary to take up again the equation of propagation in the vectorial form (1.1). In the case of a plane wave, this admits the solution:

$$\mathbf{E} = \mathbf{E}_0 \, e^{j(\omega t - \varphi)} \tag{6.1}$$

$\mathbf{E}_0$ and φ being constants. The plane of the wave (the wavefront) contains the electric field $\mathbf{E}$, that is to say, the luminous vibration, and it is normal to the light rays. In Fig. 6.1, the plane yOz is the plane of the wave. The vector $\mathbf{P}$ gives the direction of the light rays and the sense of propagation. The direction of the vector $\mathbf{P}$ is the same as that of the normal Ox to the wave. At the instant t, the electric field $\mathbf{E}$ has the components E_y and E_z. If E_{0x}, E_{0y}, and E_{0z} are the projections of $\mathbf{E}_0$ on the coordinate axes, the expression (6.1) is equivalent to the three equations:

$$E_x = 0, \qquad E_y = E_{0y} e^{j(\omega t - \varphi_y)}, \qquad E_z = E_{0z} e^{j(\omega t - \varphi_z)} \tag{6.2}$$

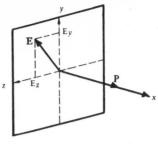

Fig. 6.1

The physical electric field is given by the real parts of the right hand sides of (6.2):

$$E_x = 0, \qquad E_y = E_{0y} \cos(\omega t - \varphi_y), \qquad E_z = E_{0z} \cos(\omega t - \varphi_z) \qquad (6.3)$$

Eliminating ωt between E_y and E_z, one obtains a curve which represents the projection of the extremity of the electric field vector on the plane yOz. This is an ellipse, and one says that, in the general case, the luminous vibration is an elliptic vibration. But the luminous vibration is emitted by the atom during a very short time, of the order of $t = 10^{-9}$ sec for even the most monochromatic type of ordinary sources. However, in order to make an observation, a very much longer time than this is needed, and during this time the atom considered, or the other atoms of the source, emit a considerable number of vibrations. These vibrations are emitted in random fashion with respect to the time, and there is no relationship between them. The quantities E_{0y}, E_{0z}, φ_y, φ_z associated with a vibration vary in a completely random fashion when one goes from one vibration to the succeeding one. There is no relationship of long duration between E_y and E_z. One has two incoherent vibrations, directed along Oy and along Oz. In order to describe the effects observed, one is led to consider time averages in the same way as before (Chapter 5). The receiver is sensitive to

$$\overline{E_y E_y^*} = \overline{E_{0y}^2}, \quad \overline{E_z E_z^*} = \overline{E_{0z}^2} \qquad (6.4)$$

This is how one may characterize the "natural light" emitted by a source. Since the choice of the directions Oy and Oz is arbitrary, natural light may be considered as produced by two perpendicular incoherent vibrations of the same intensity $E_{0y}^2 = E_{0z}^2$, because there is no preferred direction.

6.2 Polarized Light

There exist optical devices, called polarizers, which transmit only one of the two components of natural light, for example E_y. One says that the plane wave is linearly polarized. If two polarizers, placed one after the other, are oriented in such a way that the directions of the components which they transmit are perpendicular, no light is transmitted. Two *crossed* polarizers do not transmit any light. Now orient the two polarizers in an arbitrary manner with respect to one another. The first polarizer transmits the vibration OA_1 (Fig. 6.2).

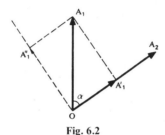

Fig. 6.2

The second polarizer transmits only the vibrations which are directed along OA_2. Decompose OA_1 into two components, one OA'_1 parallel to OA_2 and the other OA''_1 perpendicular to OA_2. The component OA''_1 is stopped, but not the component OA'_1. If $a \cos \omega t$ represents the vibration OA_1, the component OA'_1 is represented by the expression $a \cos \alpha \cos \omega t$. The amplitude transmitted by the two polarizers is $a \cos \alpha$ and the intensity (Malus's law) is:

$$I = I_0 \cos^2 \alpha \qquad (6.5)$$

If $\alpha = \pi/2$, the two polarizers are *crossed* and $I = 0$. When $\alpha = 0$, one says that the polarizers are *parallel* and the transmitted intensity is a maximum.

6.3 Young's Experiment in Polarized Light

The point source S emits monochromatic light which illuminates the two holes T_1 and T_2 (Fig. 6.3). One observes the interference fringes on the screen E_2. Place a polarizer $\mathscr{P}_1$ in front of T_1 and a polarizer $\mathscr{P}_2$ in front of T_2. Orient $\mathscr{P}_1$ and $\mathscr{P}_2$ in such a way that they transmit

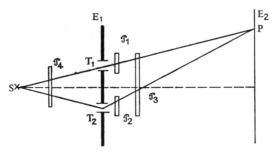

Fig. 6.3

perpendicular vibrations. The fringes disappear, because two perpendicular vibrations cannot interfere. Introduce a polarizer $\mathscr{P}_3$ as shown in Fig. 6.2. It transmits only one direction of vibration, and the vibrations emitted by T_1 and T_2 are therefore parallel after having traversed $\mathscr{P}_3$. Nevertheless, the fringes remain invisible because the two perpendicular vibrations emanating from $\mathscr{P}_1$ and $\mathscr{P}_2$ arise from two *incoherent* components of the natural light emitted by S. Place a polarizer $\mathscr{P}_4$ between S and the holes T_1, T_2. This time the two perpendicular vibrations coming from $\mathscr{P}_1$ and $\mathscr{P}_2$ are *coherent* because they originate in a unique vibration given by $\mathscr{P}_4$. The polarizer $\mathscr{P}_3$ makes these two vibrations parallel, and one observes fringes on the screen E_2.

6.4   Anisotropic Dielectric Medium. Uniaxial Medium

In an anisotropic dielectric medium, Maxwell's equations show that it is the electric displacement vector **D** which lies in the wavefront and not the electric field vector **E** (Fig. 6.4). Except for certain special

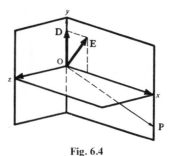

Fig. 6.4

directions, the normal Ox does not coincide with the direction of the light rays, which is given by the vector **P** normal to **E**.

Since the medium is optically anisotropic, the speed of the wave varies with the direction of the light rays. Starting from an arbitrary origin within the medium, draw in the direction of the light ray a length proportional to the speed. The locus of the extremity of this line segment is the wave surface. In the general case, the wave surface has a quite complicated form, and we shall confine our study to *uniaxial media,* such as Iceland spar or quartz. The wave surface of uniaxial media consists of two sheets: a spherical sheet and an ellipsoidal sheet. Figures 6.5 and 6.6 correspond respectively to quartz and to Iceland spar. The spherical sheet is characterized by a speed v_0 which is called the ordinary velocity. In the case of the ellipsoidal sheet, the speed varies with the direction of the ray; and its minimum value (Fig. 6.5), or maximum value (Fig. 6.6), is called the extraordinary velocity v_e. We set

$$n_o = c/v_o, \qquad n_e = c/v_e \qquad (6.6)$$

when n_o is the ordinary index and n_e is the extraordinary index and c is the speed of light in vacuum. Uniaxial media are called *birefringent* media. In the direction Ox the two speeds are equal; this direction is called the *optic axis* of a uniaxial medium.

In order to characterize an anisotropic medium, one also employs the index surface: starting from an arbitrary origin one draws along the wave normal two lengths proportional to the two indices. The locus of the extremities of these two lengths is the index surface. The

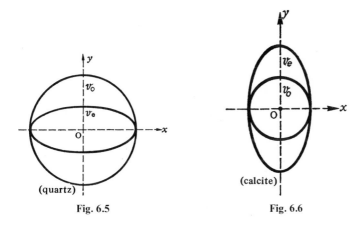

Fig. 6.5   Fig. 6.6

wave surface may be derived from the index surface by a transformation in reciprocal polars. The index surface and the wave surface have analogous forms.

6.5   Refracted Rays in a Uniaxial Medium. The Plane of Incidence Is Parallel to the Direction of the Optic Axis

Let us consider a plate with plane parallel faces cut, for example, out of Iceland spar. The optic axis is at an arbitrary angle with respect to the faces, and it is parallel to the plane of incidence. Consider rays which are incident normal to the plate and let us employ Huygens' construction. The incident ray SI cuts in the point M the spherical wave surface Σ, which has center I and which relates to the medium from which the ray is incident (air). At the point M form the plane π tangent to Σ: it intersects the entry plane of the plate at infinity. Construct the planes π_0 and π_e which are parallel to the plane π and which are tangent, respectively, at P to the spherical ordinary sheet, and at Q to the ellipsoidal extraordinary sheet. One obtains the two rays IM (ordinary) and IQ (extraordinary) which, when they emerge from the plate, are parallel.

Experiment shows that these two rays are polarized at right angles. If one calls the plane containing the normal to the entry face and the optic axis the plane of the *principal section*, the vibrations of the

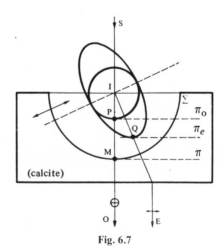

Fig. 6.7

ordinary ray O and the extraordinary ray E are, respectively, perpendicular and parallel to the plane of the principal section. One may say that the vibrations O and E originate from the two components of natural light; they are therefore *incoherent*.

If the incident rays are normal to the plate, the incident wavefront is plane and, in the uniaxial medium, one has the plane wavefronts π_o (ordinary plane wave) and π_e (extraordinary plane wave). One sees, in consequence, that the extraordinary ray IQ is not perpendicular to the plane of the wavefront.

6.6 Polarizers

A Nicol prism consists of the two halves of a rhombohedron of Iceland spar (Fig. 6.8) which have been cut apart and then glued together with Canada balsam. Figure 6.8 shows the plane of the principal section and AB represents the trace of the plane along which the crystal has been cut. As a result of the difference of the indices n_o and n_e of Iceland spar with respect to the index of the Canada balsam, the ordinary ray undergoes total reflection at I while the extraordinary ray is transmitted.

The polarizers which are most commonly used nowadays are the polaroids. These are composed of chains of molecules which are oriented parallel to one another. When the vibration is in the direction of the alignment of the molecules, the light is absorbed.

Reflection at the surface of separation of two dielectrics of indices of refraction n_1 and n_2 may also be used to produce polarized light. Let us consider a parallel bundle of rays of natural light which is reflected at the plane surface bounding a dielectric medium of index of refraction n_2 (Fig. 6.9). For a particular angle of incidence, called the

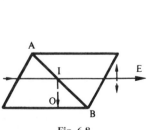

Fig. 6.8

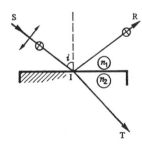

Fig. 6.9

Brewster angle of incidence, only the component of the natural light which is perpendicular to the plane of incidence is reflected. The Brewster angle is given by the relation:

$$\tan i = \frac{n_2}{n_1} = n \tag{6.7}$$

n being the index of medium (2) with respect to medium (1).

6.7   Path Difference Caused by Traversing a Birefringent Plate Parallel to the Axis

The phenomena which occur when light traverses a birefringent plate cut at an arbitrary orientation with respect to the axis are generally quite complex. They become very much simpler if the plate is cut perpendicular to the axis (Fig. 6.10) or parallel to the axis (Fig. 6.11).

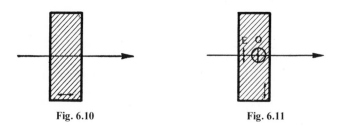

Fig. 6.10   Fig. 6.11

In the first case, if the plate is traversed by a bundle of parallel rays, normal to the plate, everything occurs as if one were dealing with an isotropic plate. This is no longer true when the rays traverse the plate at a variety of inclinations. When a plate cut parallel to the axis (Fig. 6.10) is traversed by a bundle of parallel rays, normal to the plate, the ordinary rays coincide with the extraordinary rays and there is no longer a doubling of the image as in Fig. 6.7. It is only necessary to orient the wave surface (Fig. 6.5 and 6.6) suitably with respect to the faces of the plate in order to see this at once. We shall restrict our study to the case of a plate parallel to the axis traversed by a bundle of parallel rays, normal to the plate.

In Fig. 6.11, the plane of incidence, the plane of the principal section and the plane of the figure coincide. According to the preceding paragraph, an incident vibration perpendicular to the plane of the principle section becomes an ordinary vibration in the plate. It propagates with speed v_o. An incident vibration parallel to the plane of the principal section becomes an extraordinary vibration, and propagates with speed v_e. Vibrations incident parallel or perpendicular to the plane of the principal section propagate without alteration, but their speeds are different. The direction of the axis and the directions perpendicular to it are called the *privileged directions* of the uniaxial medium.

If the incident vibration has an arbitrary orientation, one may decompose it into two components, one perpendicular to the plane of incidence and the other parallel to the plane of incidence. The first becomes the ordinary vibration and the other the extraordinary vibration. In the case of the plate studied, the ordinary ray and the extraordinary ray, though they coincide with each other, traverse different optical paths within the plate. If n_o and n_e are the indices of the plate and e is its thickness, the optical paths traversed by the ordinary ray and the extraordinary ray are respectively equal to $n_o e$ and $n_e e$. The path difference and the phase difference are

$$\Delta = (n_o - n_e)e, \qquad \varphi = \frac{2\pi(n_o - n_e)e}{\lambda} \qquad (6.8)$$

If $\Delta = p\lambda$, the birefringent plate is called a *full-wave plate* for $p = 1$.
If $\Delta = (2p + 1)\lambda/2$, the plate is called a *half-wave plate* for $p = 0$.
If $\Delta = (2p + 1)\lambda/4$, the plate is called a *quarter-wave plate* for $p = 0$.

The vibration leaving the birefringent plate is no longer linear, as was the incident vibration. In the general case, the birefringent plate transforms a linear incident vibration into an elliptic vibration.

6.8 Interference Phenomena Produced by a Birefringent Plate

We shall continue with the case of a birefringent plate with parallel faces cut parallel to the axis (Fig. 6.12). As we saw in Section 6.5, the ordinary vibration O and the extraordinary vibration E are incoherent when the plate is illuminated with natural light. Place a polarizer $\mathscr{P}_1$ in front of the birefringent plate L. Orient $\mathscr{P}_1$ in such a way that the incident vibration OA that it transmits is inclined at an angle of 45° to the plane of the principal section (Fig. 6.13). One may decompose OA into two components OA_y and OA_z of equal amplitude and in phase. In the plate, the vibration A_y becomes the extraordinary vibration, and the vibration A_z becomes the ordinary vibration. After traversing

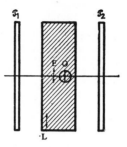

Fig. 6.12

the plate, the two vibrations OA_y and OA_z have a path difference or a phase difference given by (6.8). By virtue of the polarizer $\mathscr{P}_1$, the two vibrations are in fact coherent, but since they are perpendicular, they cannot interfere. One then places a polarizer $\mathscr{P}_2$ after the plate L. The polarizer $\mathscr{P}_2$ is also oriented at 45° to the plane of the principal section, and only passes the projections OA'_y and OA'_z of the two vibrations OA_y and OA_z on OA. In the case shown in Fig. 6.13 the two

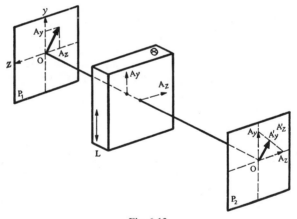

Fig. 6.13

polarizers $\mathscr{P}_1$ and $\mathscr{P}_2$ are *parallel*. The two vibrations OA'_y and OA'_z are parallel and have the same amplitude; the intensity of the light transmitted by the polarizer $\mathscr{P}_2$ is given by (2.8):

$$I_{\parallel} = I_0 \cos^2\left(\frac{\varphi}{2}\right) \tag{6.9}$$

where $I_\parallel$ means the intensity transmitted by the plate between parallel polarizers. This formula is in agreement with Fig. 6.13, because if OA_y and OA_z are in phase ($\varphi = 0$) on arriving at $\mathscr{P}_2$, the intensity I ought to be a maximum. In the case where the polarizers are crossed the arrangement of the two vibrations OA'_y and OA'_z is as shown in Fig. 6.14. They are in opposition when OA_y and OA_z are in phase and

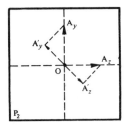

Fig. 6.14

the intensity I is zero. Between crossed polarizers the transmitted intensity $I_\perp$ is given by the expression:

$$I_\perp = I_0 \sin^2\left(\frac{\varphi}{2}\right) \tag{6.10}$$

When one orients the polarizers $\mathscr{P}_1$ and $\mathscr{P}_2$ in arbitrary fashion, the amplitudes of the interfering vibrations are no longer equal, and it becomes necessary to use formula (2.10). But we have seen that interference phenomena have their maximum visibility when the amplitudes of the interfering vibrations are equal. It is therefore necessary to arrange things so that the circumstances are those we have discussed in the preceding work: *the two polarizers are parallel or crossed, and the plate is oriented so that its privileged directions are at 45° to the directions of the vibrations which the polarizers transmit.*

The question which now presents itself is: how are we to observe the interference phenomena? The observation may be made according to the plan shown in Fig. 6.15. The birefringent plate L and the polarizers $\mathscr{P}_1$ and $\mathscr{P}_2$ are oriented in the manner we have just described. The assembly is illuminated by a parallel bundle of rays normal to the plate, and the eye is placed at the image focal point of an objective O. The focusing is carried out on the plate L at the object focal point of an objective O. If the thickness e is constant, the eye sees the plate uniformly illuminated, and the luminous intensity is given by (6.9) or (6.10).

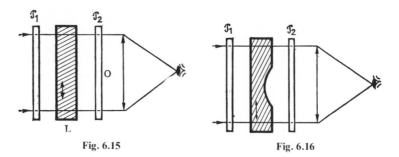

Fig. 6.15　　　　　　　　　　　　Fig. 6.16

Now let us suppose that the plate L has a variable thickness. The eye will no longer see the plate uniformly illuminated. The variations of intensity I will follow the variations in thickness.

In a region where one has $(n_o - n_e)e = p\lambda$, the intensity between crossed polarizers will be zero. One will observe a black fringe. Now move to a neighboring region where one has $(n_o - n_e)e' = (p + 1)\lambda$; then one is at the next black fringe. In passing from one black (or bright) fringe to the next black (or bright) fringe, the thickness varies by an amount equal to $\lambda/(n_o - n_e)$.

Suppose now that the plate L is illuminated by white light. If one has $(n_o - n_e)e = \lambda$ between crossed polarizers for $\lambda = 0.55$ μm, the yellow-green of the spectrum is blocked. The superposition of the other wavelengths gives a purplish tint to the transmitted light. One thus explains the appearance of colors as in Section 2.8. The same thing holds true for channeled spectra.

To sum up, one can say that there is no difference, from a fundamental viewpoint, between the interference phenomena produced by a birefringent plate and interference phenomena such as those which have been studied in Chapter 2. The only difference that exists is in the manner in which the interfering vibrations have been produced, that is to say in the method of "wave division."

6.9　Applications

The applications of interference in polarized light are very numerous. In mineralogy, the polarizing microscope allows us to study optically anisotropic crystals by examining the colors produced between polarizers. The polarizing microscope is also used in biology, where preparations often show interesting anisotropic structures.

Objects which are isotropic in normal conditions can become birefringent under the influence of stress. The stresses applied to models can then be displayed by observing the model in polarized light. This is the principle of photoelastic measurements.

One knows how to construct interferometers utilizing birefringent elements as wave divisors applicable to transparent isotropic objects. These find application in many fields, particularly in microscopy. The use of polarizing interference microscopes has been increasing, both for transmitted and for reflected light. Finally, polarized light plays a part in light modulators, in lasers, in nonlinear optics, and to some extent in so many phenomena that it is impossible to enumerate them all.

CHAPTER 7

Formation of Images. Filtering of Spatial Frequencies by an Optical Instrument

7.1 Spatially Incoherent and Spatially Coherent Objects

Incoherent objects are objects the elements of which emit vibrations that are incoherent, that is to say, that are independent of one another. Self-luminous objects are necessarily incoherent objects (the sun, the stars, various terrestrial sources of light, etc.). The same thing holds true of scattering objects illuminated by an auxiliary source, because the size of an ordinary source is such as to negate the possibility of coherence between the vibrations scattered by the various elements of the object. The case of optical instruments utilizing an auxiliary source, such as the microscope, is more complicated. In equipment such as this, in fact, the object, for example the microscopic preparation, is illuminated by vibrations coming from the same source. If the source is a point source, the illumination is coherent: the vibrations diffracted by the details of the preparation are coherent and therefore capable of interfering. When the source is not a point, the illumination is partially coherent; it is necessary to know the characteristics of the equipment in order to specify the coherence of the different points of the object.

7.2   Image of an Extended Object in Spatially Incoherent Illumination

The object is in the plane $F_0 y_0 z_0$ and the objective O gives an image of it in the plane $F yz$ (Fig. 7.1). The shaded region constitutes an arbitrary incoherent object, for example, a sketch on a sheet of paper illuminated by a lamp. We shall assume that the object is illuminated by monochromatic light. The luminosity of the object varies from one point to another, and the vibrations emitted are incoherent. Each point of the object has as its image in the plane $F yz$ a diffraction spot whose shape is determined by the shape of the aperture which delimits the bundle.

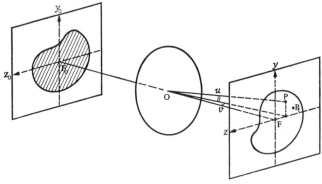

Fig. 7.1

Since all the points of the object are incoherent, it is necessary to form the sum of the intensities of all the diffraction spots in order to get the image of the extended object. We recall the notation:

amplitude diffraction pattern (image of
   a point): $\quad\quad\quad f(u, v)$
intensity diffraction pattern (image of a
   point): $\quad\quad\quad I(u, v) = |f(u, v)|^2$
intensity distribution in the geometric
   image of an object: $\quad\quad\quad O(u, v)$
intensity distribution in the image of an
   extended object: $\quad\quad\quad E(u, v)$

If the objective O is bounded by a circular aperture, the amplitude is given by the expression (3.12). We wish to calculate the luminous intensity at an arbitrary point P. Let u, v be the direction cosines of

OP, and u_1, v_1 those of a direction OP_1, P_1 being an arbitrary point of the illuminated region. If P_1 is taken to be the only source of light, one has a diffraction spot centered at P_1 which is spread out around this point. This gives rise at P to an intensity:

$$I(u_1 - u, v_1 - v) \tag{7.1}$$

Since not all the points of the geometric image have the same intensity, it is necessary to multiply the expression (7.1) by $O(u_1, v_1)$. The intensity at P is then given by:

$$E(u, v) = \iint O(u_1, v_1)I(u - u_1, v - v_1)\, du_1\, dv_1 \tag{7.2}$$

the integral being extended over the illuminated region of the geometrical image of the object. The intensity is represented by the convolution of the two functions O and I. One may write symbolically:

$$E(u, v) = O(u, v) \otimes I(u, v) \tag{7.3}$$

7.3   Image of an Extended Object in Spatially Coherent Illumination

In the case of an extended object in coherent illumination, one has the situation of Fig. 7.2. The object is located at A and it is illuminated by a parallel bundle. The objective O forms at A' an image of the object

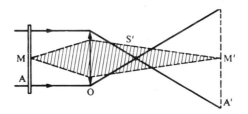

Fig. 7.2

A. An arbitrary point M of the object diffracts light which is concentrated by O to form the image M'. The image M' is a diffraction spot which is characteristic of the objective O. The image of the extended object at A' is the result of summing the amplitudes of all the diffraction phenomena produced by all the points of the object. The problem is treated in the same way as before, but it is necessary to take

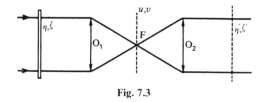

Fig. 7.3

the sums of the amplitudes and not the intensities. The relation (7.3) is replaced by a convolution on the amplitudes.

In the case of coherent illumination, one may make the calculation by means of a double diffraction (Fig. 7.3). The object is located in the plane η, ζ and it is illuminated by parallel light. An objective O_1 placed after the object gives in its focal plane at F the diffraction phenomena which are characteristic of the object. A second objective O_2 gives at η', ζ' an image of the object. One may therefore calculate the image in two steps:

1. applying the Fourier transformation (formulas (3.5) and (3.6)), one calculates the diffraction phenomena at F.

2. one goes from the diffraction phenomena at F to the image at η', ζ' by means of a second Fourier transformation.

As a natural consequence of the finite apertures of the objectives O_1 and O_2, all the vibrations diffracted by the object are not recombined in the image, and this latter cannot be rigorously identical to the object. In fact, the preceding calculation brings in integrals which are not extended from $-\infty$ to $+\infty$.

7.4   Transfer Function of an Optical Instrument. Incoherent Object

The concept of resolving power is based on the possibility of using an optical instrument to separate two luminous points of the same intensity whose images are diffraction spots. If the distance between the two diffraction spots is sufficiently large, the eye sees two distinct images. When the two diffraction spots are too close, the eye essentially sees only one image. The resolving power is given by the angular (or linear) distance of two luminous points when the eye can no longer distinguish the presence of two images.

This notion of resolving power is very subjective, and does not give an accurate representation of the performance of an optical instrument.

Nowadays one prefers to characterize the qualities of an instrument by its *transfer function*. The optical instrument is considered as a filter which does not transmit all the details of the object in a uniform fashion to the image. It is, in fact, evident that large details are better transmitted than fine details. It is just this mechanism of transmission which it is important to state precisely, in order to characterize the quality of an instrument. An object possessing the whole range of details, from the finest to the largest, is therefore necessary for this study. This is the case for the object we are going to describe.

Consider an assembly of parallel equidistant strips which are alternately white and black. They are drawn for example on a sheet of paper and illuminated by monochromatic light. Such an object is an incoherent object. The bright strips have a width Z_1 which is small compared with the distance Z_0 which separates them (Fig. 7.4). If

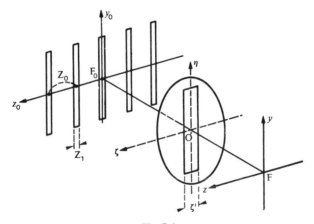

Fig. 7.4

the bright strips were slits, one would have an actual grating. In Fig. 7.4 the periodic object is in the plane $F_0 y_0 z_0$, and the objective O gives an image of it in the plane $F y z$. It is the mechanism of formation of this image that we are going to study. For simplicity's sake, we may assume that the magnification of the objective is equal to 1. The intensity profile of the object or of its geometric image (Fig. 7.5) may be represented by the expression:

$$O(z) = 1 + 2\cos\frac{2\pi z}{Z_0} + 2\cos\frac{2(2\pi z)}{Z_0} + \cdots \qquad (7.4)$$

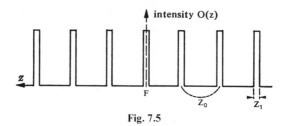

Fig. 7.5

If we designate as *spatial frequency* the reciprocal of the *step* of a sinusoid, the object $O(z)$ is composed of an infinity of sinusoids having spatial frequencies zero, $1/Z_0$, $2/Z_0$, etc. To the sinusoids of low frequencies $1/Z_0, 2/Z_0, \ldots$ correspond the large details of which we have spoken. To the sinusoids of high frequency there correspond the fine details. Thus, despite what might be one's first impression, the object shown in Fig. 7.5 contains in itself details whose spatial frequencies range from $1/Z_0$ to infinity.

Let us study how the spatial frequencies are transmitted by the objective O. It is the relation (7.3) which gives the solution of the problem. It is known that, in this relation, the three quantities represent intensities. Take the Fourier transform of the two sides:

F.T.(image of the object) = F.T.(geometric image of the object)

$\times$ F.T.(image of a point)

(7.5)

The Fourier transform of the image of an incoherent object is equal to the product of the transform of the geometric image of the object by the transform of the image of an isolated luminous point.

Now let us look for the transform of the intensity distribution of the image of a point. It is known (Section 3.3) that the transform of $|f(v)|^2$ is given by the autocorrelation function of $F(\zeta)$, that is to say by the autocorrelation function of the amplitude distribution over the pupil. Suppose, for simplicity, that the objective forming the image of the object were bounded by a slit of width ζ'. The autocorrelation function is represented by the curve of Fig. 7.6. If R is the distance of the image from the objective, one sets:

$$\alpha = \frac{\zeta'}{R} \qquad (7.6)$$

where α is the aperture of the objective which forms the image. This figure corresponds to the second factor of the right-hand side of (7.3).

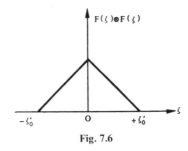

Fig. 7.6

Now let us determine the transform of the geometric image of the object. We have seen that the transform of a periodic function of the type shown in Fig. 7.5 is given by $\mathrm{Comb}(v/v_0)$, where $v_0 = \lambda/Z_0$. It is unnecessary to bring in here the factor $\mathrm{sinc}(\pi v Z_1/\lambda)$, since Z_1 is assumed small compared with Z_0. It is worth noting that in the present case the transform $\mathrm{Comb}(v/v_0)$ does not have a real existence; it is purely a mathematical representation. The incoherent object which we are considering has, in intensity, the same profile as that of a grating in coherent light; and consequently the two profiles have the same Fourier transform. This transform is observable with the grating, and is not observable with the incoherent object.

Knowing the transform of the image of a point and the transform of the object, one can determine the transform of the image of the object. We show in Fig. 7.7 the transforms of the object and of the image of a point. In order to determine the transform of the image, it suffices to form the product of the two transforms. In the example shown in Fig. 7.7, one sees that the spatial frequency $1/Z_0$ is attenuated but it is transmitted by the objective O. On the other hand, all the other spatial frequencies are eliminated. The optical instrument behaves therefore like a low-pass filter.

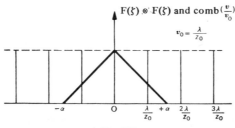

Fig. 7.7

Now suppose that the amplitude of the spatial frequency $1/Z_0$ is multiplied by a factor $M < 1$. In accordance with (7.4) the image will be represented by the expression:

$$E(z) = 1 + 2M \cos \frac{2\pi z}{Z_0}, \qquad M < 1 \qquad (7.7)$$

In the neighborhood of the limit of resolution, the image is rigorously sinusoidal. The limiting case is attained when the lowest spatial frequency is such that $\lambda/Z_0 = \alpha$. The limiting spatial frequency transmitted by the instrument is:

$$\frac{1}{Z_0} = \frac{\alpha}{\lambda} \qquad (7.8)$$

If the spatial frequency $1/Z_0$ is not transmitted, there is no longer any image. The more spatial frequencies are transmitted, the more nearly the structure of the image approximates that of the object. If all the spatial frequencies were transmitted, that is to say, if the aperture α of the objective were infinite, the image would be identical to the object. The mechanism of transmission of frequencies is therefore essential in the formation of images. We see that the curve of Fig. 7.6 plays a fundamental role in this mechanism. It is this curve which determines the transmission of spatial frequencies. The function which represents this curve, that is to say, the transform of the image of an isolated point (in intensity) is called the *transfer function* of the instrument. An instrument is characterized by its transfer function and no longer by the subjective notion of "resolving power." The transfer function gives all the information needed, because it shows how the different spatial frequencies are transmitted. We have shown in Fig. 7.6 the transfer function when the objective is bounded by a slit. Evidently the form of this function depends on the shape of the aperture and the aberrations of the instrument.

7.5   Transfer Function in Coherent Illumination

One may also study the transmission of spatial frequencies by an optical instrument in coherent light. This is a simpler problem than the preceding one. The periodic object is now an actual grating in coherent light. One obtains an image of the grating which becomes better and better as the number of spectra transmitted by the instrument increases. The calculation is carried out very easily, starting from the basic facts of Chapter 4. We shall study a case of filtering in coherent light in Chapter 10 (optical processing of information).

7.6   Phase Contrast and Strioscopy

Objects characterized by variations of index or of thickness are called "phase objects." These objects, which are taken to be perfectly transparent, show no contrast with the background and are invisible by ordinary methods. Consider a plane parallel plate of glass which contains a small region A of thickness e and of index n different from the index n' of the rest of the plate (Fig. 7.8). The path difference between the ray (1) which passes through the region A and an arbitrary ray (2) which passes alongside is $\Delta = (n - n')e$. The object A is a phase object which is characterized by the path difference Δ. In everything that follows, we shall suppose Δ and $\varphi = 2\pi\Delta/\lambda$ to be small. Reflecting objects also constitute phase objects. Let A be such an object (Fig. 7.9): irregularities resulting from the polishing process modify the optical paths traversed by the rays (1) and (2) reflected by the surface A. With respect to a plane reference surface Σ, the optical path difference is equal to $2e$, where e is the difference in thickness of the object A in the regions M_1 and M_2.

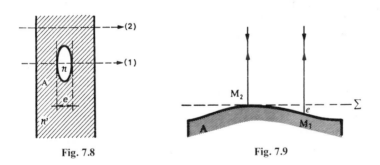

Fig. 7.8   Fig. 7.9

Objects of the type characterized solely by variations of optical path, and not by variations of amplitude, are not visible by ordinary methods of observation. The method of phase contrast makes it possible to change the phase variations of the object into amplitude variations in the image. The phase objects thus become visible. Consider, for example, the object A of Fig. 7.8. It is illuminated by a parallel

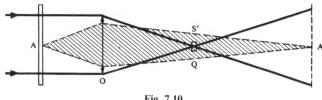

Fig. 7.10

bundle of light (Fig. 7.10). After traversing the objective O, the bundle is brought to a focus at the focal point S' of the objective O, and then spreads out into the image A' of A. We shall call this bundle the direct light bundle. The phase detail A diffracts some light which is focused on the image A' by the objective O. The image A' results from the interference of the direct light, which produces a coherent background, with the diffracted light. At a general point of the object, the complex amplitude is proportional to $e^{j\varphi}$, where $\varphi = (2\pi/\lambda)(n - n')e$. Since φ is small, we have:

$$e^{j\varphi} \cong 1 + j\varphi \qquad (7.9)$$

The first term on the right-hand side represents the direct light, and the second term $j\varphi$ the diffracted light. In fact if $n = n'$, that is to say if the object disappears, $\varphi = 0$ and there remains only the first term. Now place at S' a small very thin plate Q. All the direct light passes through Q, but since the diffracted light is spread out in this region, one may neglect the small fraction of the diffracted light which passes through Q. Give the plate Q an optical thickness such that the vibrations of direct light which pass through it are retarded by $\lambda/4$ (phase difference $\pi/2$) with respect to the vibrations of diffracted light which pass to one side of Q. One then has in the image A' the amplitude:

$$e^{-j\pi/2} + j\varphi = e^{-j\pi/2}(1 - \varphi) \qquad (7.10)$$

and the intensity is:

$$I_A = 1 - 2\varphi \qquad (7.11)$$

since φ^2 may be neglected compared to φ. Outside of the image A' one has $I = 1$. The image A' then displays a contrast equal to φ. Instead of retarding the direct vibrations, one may advance them by taking an optical thickness $3\lambda/4$ which is equivalent to an advance of $\lambda/4$ with respect to the diffracted vibrations. The plate Q is called a *phase plate*. One may increase the sensitivity by making the phase plate absorbing.

Suppose that it diminishes the intensity of the direct light by a factor N. Then one has:

$$I_A = \left(\frac{1}{\sqrt{N}} - \varphi\right)^2 = \frac{1}{N}(1 - 2\varphi\sqrt{N}) \qquad (7.12)$$

and in the rest of the field $I = 1/N$. From this we see that the contrast is:

$$\gamma = \varphi\sqrt{N} \qquad (7.13)$$

Thus the contrast is multiplied by $\sqrt{N}$. In principle, if $N = 2500$ one might observe path differences of 10 Å with a contrast of 0.5, which is a good contrast.

At the present time, the method of phase contrast is employed primarily in microscopy. Phase contrast microscopes are used in most biology laboratories because they permit the study of living objects without coloring them and thus without killing them.

Replace the phase plate Q by an opaque plate. The amplitude at A' is given by $j\varphi$ and it is zero outside the image A', because all the direct light is blocked off by the opaque screen Q.

The intensity is then φ^2, and it is zero in the rest of the field. The contrast is a maximum and is always equal to 1. This is the method of strioscopy or the method of the dark field. Unfortunately, if φ is small then φ^2 is that much the smaller and the images have a very low brightness. Besides, even the most minor of optical defects, or grains of dust, can diffract a lot of light and veil the image. And finally, the images of objects which have a complicated structure are difficult to interpret.

CHAPTER 8

Holography

8.1 Introduction

In 1948 D. Gabor described a novel concept he called "Holography" which made it possible to reconstruct the image of an object starting from the diffraction pattern produced by the object. In this method, a part of the light emitted by a monochromatic light source illuminates the object and the remaining part, which constitutes a "coherent background," falls directly on the photographic plate. The photographic plate records the variations in intensity caused by the interference of the coherent background and the light scattered by the object on to the plate. Once it has been developed, the photographic plate constitutes the "hologram."

In contrast to what happens with an ordinary photograph, one sees no image of the object on the photographic plate. But if one illuminates the "hologram" with monochromatic light, one may see, through the hologram, a three-dimensional image of the object. Obtaining a coherent field was a very difficult problem in 1948 because the sources known at that time were not very monochromatic. For an object of moderate size, it is in fact necessary that the coherence length of the light utilized shall be sufficiently long so that the interference phenomena between the coherent background and the light scattered by the object may be produced in the plane of the photographic plate. It was necessary to wait for the invention of the laser before holography could take on its full development. It was the physicists at the University of Michigan, and in particular F. N. Leith and J. Upatnieks, who conceived, in 1962, the decisive improvements on the method of D. Gabor. Before describing the principle of holography, it will be necessary to remind the reader of the basic properties of photographic recording.

8.2   Amplitude Transmitted by a Photographic Plate after Development

Expose a photographic plate. After development, illuminate the negative. The transmission factor of the negative is:

$$T = \frac{\text{transmitted intensity}}{\text{incident intensity}} \tag{8.1}$$

which is always less than unity. The logarithm of $1/T$ is the density D of the negative:

$$D = \log\left(\frac{1}{T}\right) \tag{8.2}$$

Denoting the amplitude transmission coefficient by t, one has:

$$T = t^2, \quad D = \log\left(\frac{1}{t^2}\right) \tag{8.3}$$

If I is the intensity falling on the plate during exposure at an arbitrary point, and if τ is the exposure time, the plate receives the energy $W = I\tau$. One gives the name *blackening curve* or *characteristic curve of the emulsion* to the curve giving the variation of the density D (on the negative) as a function of the energy W received by the plate (Fig. 8.1).

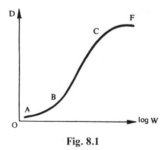

Fig. 8.1

The curve has a straightline portion described as the normal exposure region and two other regions, one *AB* corresponding to underexposure and the other *CF* corresponding to overexposure. Let γ be the slope of the straightline portion; then we have in this region:

$$D = \gamma \log\left(\frac{W}{W_0}\right) \tag{8.4}$$

where W_0 is a constant. In holography, the significant relationship is that which exists between the *amplitude* transmitted by the negative and the energy W received by the plate. The representative curve of the function $t = f(W)$ is shown in Fig. 8.2. One may interpret this

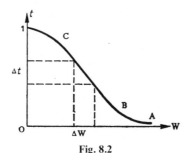

Fig. 8.2

curve by saying that if the amplitude falling on the negative is equal to 1, the amplitude emerging from the negative is equal to t. The curve $t = f(W)$ also has a straightline portion *BC* which plays an important role. It is worthy of remark that the straightline portion of the curve $t = f(W)$ corresponds to the region of underexposure of Fig. 8.1. In the case where one makes a single exposure, it is not necessary to take the exposure time into account, and one may study the phenomena with the help of the curve $t = f(I)$ relating t to I. The curve of Fig. 8.2 shows that if the variations ΔW of the energy received by the plate lie in the straightline region, the corresponding variations Δt of the amplitude transmitted by the negative are proportional to ΔW.

8.3   The Gabor Hologram. Case Where the Object Is a Simple Luminous Point

Let A_0 be a monochromatic point source (Fig. 8.3) which illuminates the plane η, ζ containing the photographic plate H. The luminous point A_0 constitutes the "object." In addition to the spherical

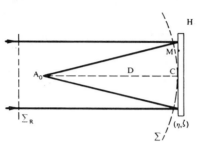

Fig. 8.3

wave Σ emanating from A_0, the plate receives a plane wave Σ_R which is coherent with Σ. This implies that both Σ and Σ_R originate from the same light source, which is not shown in Fig. 8.3. For simplicity, let us assume that Σ and Σ_R are in phase at C. For the moment, we shall not concern ourselves with the way in which these two waves are produced starting with the initial source; this is a technical matter about which we will go into details later on.

At a general point M located at a distance ρ from the point C the spherical wave Σ gives rise to the complex amplitude $F(\eta, \zeta) = b \exp(-jK\rho^2/2D)$, where $D = \overline{A_0 C}$ and $K = 2\pi/\lambda$. To a good approximation $\rho^2/2D$ represents the gap between Σ and the plane H. The coherent wave Σ_R has a constant amplitude a in the plane of the plate (a is real). At M the total amplitude is $a + F(\eta, \zeta)$ from which the intensity after some simplification is:

$$I = (a + F)(a + F^*) = a^2 + |F|^2 + aF + aF^* \qquad (8.5)$$

The variations of intensity over the plate are due to the interference of the two waves Σ and Σ_R. One then finds circular rings centered at C which are analogous to Newton's rings. Since the intensity I may be written as $(a^2 - b^2) + 4ab \cos^2(K\rho/4D)$, one sees that the fringes have maximum contrast when $a = b$, that is to say, when the two waves have the same amplitude.

Let τ be the exposure time. The plate receives energy $W = I\tau$, and after development the negative transmits an amplitude which is proportional to W if one is operating on the linear portion of the curve $t = f(W)$ (Fig. 8.2). In order for this to come about, it is necessary that W shall not depart too much from a mean value W_0, that is to say, that the interference fringes shall not have too high a contrast ($a \neq b$).

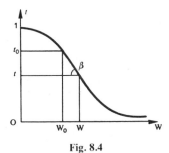

Fig. 8.4

Setting up these conditions, the amplitude transmitted by the negative may be written (Fig, 8.4):

$$t = t_0 - \beta(W - W_0) \tag{8.6}$$

Set $W_0 = \tau a^2$ and $\beta' = \beta\tau$; using (8.5) we find:

$$t = t_0 - \beta'(|F|^2 + aF + aF^*) \tag{8.7}$$

The negative thus obtained is the hologram of the object, which in this case is the luminous point A_0.

8.4   Reconstruction of the Image of a Luminous Point

Illuminate the hologram H with a uniform plane wave, parallel to the plane of the hologram and of amplitude equal to unity (Fig. 8.5). The amplitude transmitted is given by the expression (8.7). Let us consider the terms t_0, $\beta'|F|^2$, $\beta'aF = \beta'ab\exp(-jK\rho^2/2D)$ and $\beta'aF^* = \beta'ab\exp(jK\rho^2/2D)$. The first corresponds to a constant amplitude. As regards this term, everything happens as if the hologram were uniform. It thus represents a plane wave directly transmitted by the hologram.

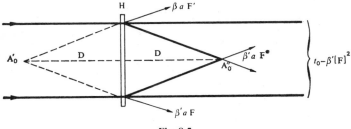

Fig. 8.5

The same thing is true of the term $\beta'|F|^2$ because the intensity $|F|^2$ produced in the plane H by A_0 is practically constant. The term $\beta'ab\exp(-jK\rho^2/2D)$ is identical, to within a factor $\beta'a$, to the amplitude produced by A_0 during the recording of the hologram. But $b\exp(-jK\rho^2/2D)$ represents, as we have seen, a spherical diverging wave. It seems to come from the point A_0' located at a distance D from the hologram. *The hologram thus reconstructs at A_0' a virtual image of A_0.* As to the term $\beta'ab\exp(jK\rho^2/2D)$, this corresponds to a convergent spherical wave which gives a real image of A_0 at a point A_0'' located at a distance D from H. All these results were in fact already found in Section 3.13, because the hologram of a luminous point is nothing more than the Fresnel zone plate studied in Section 3.13.

To sum up, if one illuminates the hologram H with a parallel bundle of rays normal to H, one observes: (1) a directly transmitted bundle and (2) a virtual image A_0' and a real image A_0'' symmetrically placed with respect to the hologram.

It is important to note that the hologram reconstructs the images A_0' and A_0'' only if the photographic plate is capable of recording all the details of the interference pattern of Σ and Σ_R, which here consists of the rings. It is in fact the diffraction produced by this system of rings which gives the images A_0' and A_0''. Since the rings come closer together in proportion as one moves away from their center, the photographic plate must have sufficient resolving power to record them. The whole mechanism of the formation of images in holography depends on this fundamental point.

8.5   Remark

If the two waves Σ and Σ_R are not in phase at C as we assumed, the spherical wave emitted by A_0 may be written $b\exp(-jK\rho^2/2D) \times \exp(j\varphi)$, where φ is the phase difference at C between Σ and Σ_R. The hologram reconstructs the two spherical waves

$$\beta'ab\exp(-jK\rho^2/2D)\exp(j\varphi)$$

and

$$\beta'ab\exp(jK\rho^2/2D)\exp(-j\varphi),$$

that is to say, the two images A_0' and A_0'' are still at the distance D from the hologram. There is nothing that has changed.

8.6 Utilization of a Coherent Background Inclined at an Angle. The Hologram of Leith and Upatnieks

Figure 8.5 shows that, with the eye located behind the hologram (to the right of H) the virtual image A'_0 will be seen, but it will be overlaid with the conjugate image A''_0. This will not be very important in the case of a point object, but can become very distracting with an extended object. In order to achieve a good separation of the bundles, one makes the recording using the experimental arrangement shown in Fig. 8.6. The coherent background is provided by a bundle of

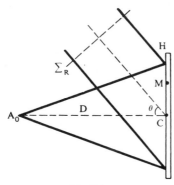

Fig. 8.6

parallel rays making an angle θ with the normal to the plate H. Taking the origin of phases at the point C, the coherent wave Σ_R produces in the plane H the amplitude $ae^{-jK\theta\zeta}$. We assume that the axis $c\zeta$ is situated in the plane SA_0C which coincides with the plane of the figure. The spherical wave emitted by A_0 gives at a general point M of the hologram H the amplitude

$$F(\eta, \zeta) = b \exp[-jK(\eta^2 + \zeta^2)/2D]$$

from which the intensity, after some simplification, is:

$$I = (ae^{-jK\theta\zeta} + F)(ae^{jK\theta\zeta} + F^*) = a^2 + |F|^2 + aFe^{jK\theta\zeta} + aF^*e^{-jK\theta\zeta}$$

$$(8.8)$$

After development, if one illuminates H with a plane wave parallel to H, the same calculation as before gives for the transmitted amplitude:

$$t = t_0 - \beta'[|F|^2 + aFe^{jK\theta\zeta} + aF^*e^{-jK\theta\zeta}] \qquad (8.9)$$

Now illuminate the hologram by a plane wave $e^{-jK\theta\zeta}$ whose orientation, with respect to the hologram, is the same as that of the reference wave used during the recording process. The amplitude transmitted is:

$$te^{-jK\theta\zeta} = (t_0 - \beta'|F|^2)e^{-jK\theta\zeta} - \beta'aF - \beta'aF^*e^{-jK2\theta\zeta} \quad (8.10)$$

As we have seen, the first term on the right-hand side of (8.10) corresponds to the plane wave directly transmitted by the hologram. The only difference arises from the presence of the factor $e^{-jK\theta\zeta}$ due to the inclination θ of the wave. The term

$$\beta'aF = \beta'ab \exp[-jK(\eta^2 + \zeta^2)/2D]$$

reconstructs a divergent spherical wave which appears to originate at the point A'_0 which is a virtual image of A_0. The last term $aF^*e^{-jK2\theta\zeta}$ reconstructs at A''_0 a real image of A_0. The presence of the factor 2θ in the exponential indicates that the image A''_0 is found in a direction making an angle 2θ with the normal to H. And now, when one positions one's eye as indicated in Fig. 8.7, one sees the virtual image A'_0 without being distracted by the presence of other luminous bundles.

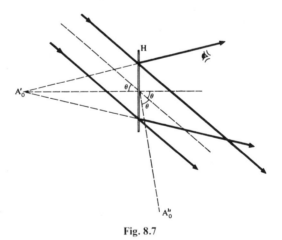

Fig. 8.7

One may remember the following result: if the hologram is illuminated in the same way as during the recording process, the virtual image A'_0 occupies the same position with respect to H as the object A_0.

8.7 Case of a General Object

A general object may be considered to be composed of a large number of luminous points. If the object is illuminated by a mono-chromatic point source, the points of the source behave like point sources of determinate amplitudes and phases. Record the hologram as shown in Fig. 8.8. A general diffusing or scattering object is illumi-nated by a point source *S* at infinity (monochromatic light). The photographic plate *H* receives the light scattered by the object and, thanks to the auxiliary mirror *m*, it also receives a bundle which proceeds directly from the source (coherent background). This is the same schematic arrangement as that of Fig. 8.6, but the point object is replaced by a general three-dimensional object *A*.

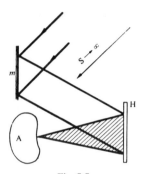

Fig. 8.8

Besides this, in the schematic of Fig. 8.8 one sees the whole experi-mental layout. If the coherence length of the vibrations is sufficiently long, the light scattered by the different points of the object will interfere with the coherent background. Here the interference pattern is extremely complicated, and it is necessary that the photographic plate be capable of recording all the details. The hologram thus reconstructs all the points of the object, i.e., the object itself. In fact, if $F(\eta, \zeta)$ now represents the amplitude falling on *H* from the *whole* object, the illumination at a general point *P* of *H* is given by (8.8). Illuminate *H* by a wave $e^{-jK\theta}$ whose orientation, with respect to the hologram, is the same as that of the reference wave utilized during the recording of the hologram: the transmitted amplitude is given by (8.10). To within a constant factor $\beta'a$ the term $\beta'aF$ corresponds to

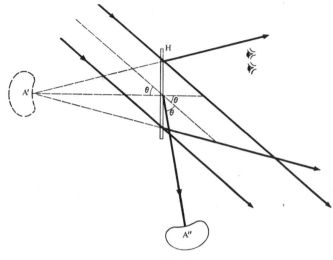

Fig. 8.9

an amplitude distribution in the hologram which is exactly that produced by the object at the moment of recording. Thus from this term the hologram reconstructs a virtual image A' of the object (Fig. 8.9). This is the same process as before but the spherical wave $\exp[-jK(\eta^2 + \zeta^2)/2D]$ is replaced by a very complicated wave $F(\eta, \zeta)$. Instead of reconstructing a spherical wave which appears to come from the virtual image of a point object, the hologram reconstitutes the wave $F(\eta, \zeta)$ that seems to come from the different points of a virtual image of the object. The image A' is identical with the object and occupies the same position with respect to H as the object itself. Through the hologram H one sees a three-dimensional image of the object. One sees a relief which is real, and not subjective as in stereoscopic vision.

The term $aF^* \exp(-jK2\theta\zeta)$ corresponds to a real image A'' of the object, an image which may be photographed directly. Nevertheless, for holograms of normal dimensions, the depth of field is so small that it is practically impossible to obtain a photograph, unless one utilizes only a very small portion of the hologram. This effect does not occur when one observes the virtual image because it is the pupil of the eye, which is always very small, that acts as the diaphragm. Besides this, the effects of parallax are not the same as when one compares the real image and the object itself.

8.8   Fourier Holograms

Consider the experimental setup shown in Fig. 8.10. A bundle of parallel rays illuminates a plane object A and a prism V. The bundle deviated by V covers the photographic plate H_F. This is the reference bundle. The bundle that passes through A is collected by an objective O whose focal plane coincides with A. Consider, to begin with, the Fourier hologram of a single point A_0 of the object A. Let us calculate the amplitude at a general point P of H_F. Take a system of axes Fyz in the plane H_F, the axis Fz being located in the plane of the figure. For simplicity, we shall assume that the point A_0 is on the axis OF, and that the ray of the reference bundle passing through F is in the plane of the figure. If θ is the inclination of the reference bundle with respect to the normal to H_F, the amplitude at P produced by the reference wave is $ae^{-jK\theta z}$.

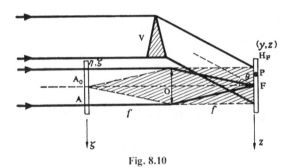

Fig. 8.10

The light diffracted by A_0 gives, after traversing O, a plane wave b, from which the total amplitude at P is:

$$ae^{-jK\theta z} + b \tag{8.11}$$

and the intensity is:

$$I = a^2 + b^2 + abe^{-jK\theta z} + abe^{jK\theta z} \tag{8.12}$$

One may also write $I = a^2 + b^2 + 2ab \cos K\theta z$ and consequently the waves from point A_0 give rise, by interference with the reference bundle, to a system of sinusoidal fringes on H_F.

In order to observe the images of such a hologram, one proceeds as shown in Fig. 8.11. The hologram H_F is illuminated normally by a parallel bundle, and the images are observed in the focal plane of an

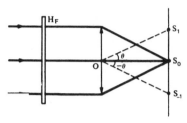

Fig. 8.11

objective O placed behind the hologram H_F. According to (8.6) the amplitude transmitted is:

$$t = t_0 - \beta'[b^2 + abe^{-jK\theta z} + abe^{jK\theta z}] \qquad (8.13)$$

The hologram gives rise to a directly transmitted plane wave $t_0 - \beta'b^2$ and two diffracted plane waves $abe^{-jK\theta z}$ and $abe^{jK\theta z}$ in the directions $v = \pm\theta_0$. These three waves give rise in the focal plane of the objective O to three point images S_0, S_1, and S_{-1}. These are the spectra of a sinusoidal grating consisting of the Fourier hologram. Either one of the images S_1 and S_{-1} may be considered as the image of the point A_0. One may also say that the images S_0, S_1, and S_{-1} are given by the transform of t in (8.13); this includes the transform of $t_0 - \beta'b^2$ and the transform of $2\,ab\cos K\theta z$, to which correspond S_1 and S_{-1}.

Instead of taking a point of the object, let us study what happens as regards the whole object plane A. It is necessary to sum up the amplitudes produced at P by all the points of the object A. The amplitude at the point y, z of the hologram is then:

$$ae^{-jK\theta z} + \Sigma be^{-jK(\eta y + \zeta z)/f} \qquad (8.14)$$

where η and ζ are the coordinates of a general point of A. Each point of the object gives rise, by interference with the reference bundle, to a sinusoid of determinate orientation and "step." For each point of the object, the hologram gives rise to two point images S_1 and S_{-1}. Finally, one obtains two images of the object which are symmetric with respect to S_0. If the object is represented by the Chinese character 木, the two images have the orientation shown in Fig. 8.12. At S_0 one has a very brilliant spot corresponding to the directly transmitted wave.

The following remark may be interesting in view of the application to the problems of recognition of shapes (Chapter 10). The total amplitude transmitted to P by all the points of the object A is nothing

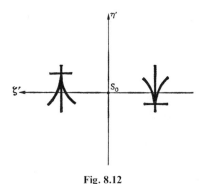

Fig. 8.12

other than the Fourier transform of the amplitude distribution over the object A. Representing this transform by $f(y, z)$, the expression (8.14) may be written:

$$ae^{-jK\theta z} + f(y, z) \tag{8.15}$$

from which the intensity is:

$$I = a^2 + |f|^2 + ae^{jK\theta z} \cdot f + ae^{-jK\theta z} \cdot f^* \tag{8.16}$$

and the amplitude transmitted by the hologram is:

$$t = t_0 - \beta'[|f|^2 + ae^{jK\theta z} \cdot f + ae^{-jK\theta z} \cdot f^*] \tag{8.17}$$

The last term gives rise to a transmitted amplitude proportional to the expression which is the conjugate of the Fourier transform of the object A. The corresponding wave is oriented in the direction of the reference bundle which was used to record the hologram.

8.9 Phase Holograms

The holograms which we have been considering have a structure which is characterized by variations of blackening. One may transform these into phase holograms by bathing them in a suitable solution (bleached hologram). The thickness of the emulsion is reduced in proportion as the blackening is greater. The hologram becomes perfectly transparent and is characterized only by variations in the thickness of the emulsion. These phase holograms are evidently much more luminous, and they also give rise to good images.

8.10   Reflection Holograms and Holograms in Color

One may create reflection holograms by using the principle of
stationary waves. The photographic plate is illuminated on one side
by the object, and on the other side by the coherent bundle. Stationary
waves are formed in the emulsion. After development, one illuminates
the hologram with *white light* and one observes by reflection. The
hologram reflects only light of the same sort as that to which it has
been exposed, and one sees an image of the object in a single color.
This is the principle of interference filters, of which we shall speak in
Section 9.10. If the hologram had been recorded in three wavelengths
(red, green and blue radiation), at the time of reconstruction and when
it is illuminated in white light, the hologram gives an image of the
object in colors and in three dimensions.

8.11   Computer-Generated Holograms

From the very fact that the photographic plate on which the
hologram is recorded contains only variation in blackening, it ought
to be possible to create such variations artificially, and consequently
to synthesize a hologram.

Synthetic holograms allow us to generalize the possibilities of
classical holography. The object from which one has constructed a
hologram no longer needs to have a real existence. One may define a
general object by giving the coordinates and the intensities of its
points; the hologram obtained allows one to visualize the imagined
object in three dimensions. It is possible to construct mathematical
objects in space, and to represent objects in the course of manufacture
without its being necessary to construct models.

In other fields, synthetic holograms are capable of reconstructing
wave surfaces of predetermined forms, for example, aspherical wave
surfaces which may be used for tests for the control of instruments by
interference methods. They equally well provide original solutions to
the problems of optical filtering, of information storage, and of a more
general method for the optical processing of information.

The operations are as follows:

(1)   Knowing the object for which one wants to create the hologram, one calculates the complex amplitude which it produces in a
plane located at a certain distance, a plane which will be that of the
hologram. The calculations are carried out by a computer.

(2)   The complex amplitude thus calculated is coded by the computer in such a way as to transform it into a real positive function. For example the computer adds to the complex amplitude produced by the object a general complex amplitude which plays the role of a coherent background. The resultant intensity is, in this case, the real positive function.

(3)   A suitable system controlled by the computer permits us to create on a plane a plot which reproduces the values of this function. This system may be a printer, a cathode-ray tube, etc.

(4)   The plot thus obtained is photographed, and the negative constitutes the synthesized hologram. In order that the hologram may diffract the light effectively, it is necessary that the plot have a sufficiently fine structure, and it is generally necessary to photograph the tracing with a magnification much less than unity.

The complex amplitude produced by the object in the plane of the hologram may be coded by the computer in several ways:

(a)   By a binary system, in which one depends only on the position and the dimensions of the small black and white signals to represent the phase and the amplitude of the object wave in the plane of the hologram. One obtains a structure which does not resemble that of an experimental hologram, but which produces in practice the same diffraction.

(b)   By adding to the complex amplitude produced by the object a general complex amplitude (coherent background). The computer calculates the resultant intensity, the variations of which represent the structure of the hologram. The method is the same as in the case of the real experiment.

(c)   By multiplying the phase of the wave produced by the object in the plane of the hologram by a function which always returns the phase to zero each time it passes through the value 2π. Such a profile may be obtained in practice from the computer calculations, and the hologram thus obtained gives only a single image of the object. A hologram of this type is called a "kinoform."

Computers do not have infinite capabilities, and the preceding calculations can be carried out for only a finite number of points. The operations are therefore digitalized. The object is described by the coordinates of a finite number of its points and by the amplitudes which they emit. The calculation of the complex amplitude in the plane of the hologram is carried out also for a finite number of points, a number which is at least equal to that of the points of the object. Take the example of a Fourier hologram: the object, the spectrum of the hologram, is an object of finite dimensions. One has, therefore, a limited spectrum, and the complex amplitude in the plane of the hologram ought to be capable of being calculated exactly from discrete values distributed over a group of equidistant points. The complex amplitude at any point whatever is then obtained by means of a very simple interpolation formula.

Finally, one may consider three principal types of synthetic holograms: binary holograms, holograms at several levels of intensity, and the holograms called "kinoforms." The two diagrams which follow

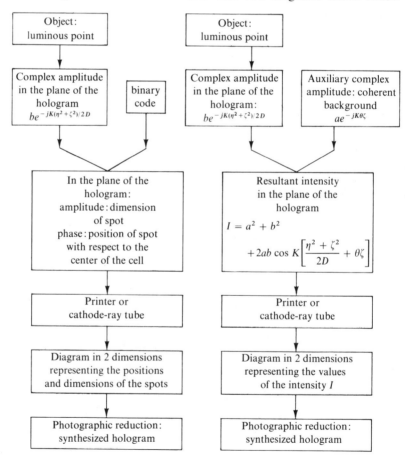

show the steps of the operations in the first two cases. For simplicity, we have taken the very simple example of a point object.

8.12   Binary Holograms of the Fourier Type

As an example of the synthesis of a hologram by means of the computer, we shall give the principle of binary holograms. It is an interesting example, because it brings in the most basic concepts of interference and diffraction.

In an ordinary hologram, the phases are recorded with the aid of a coherent background. There is interference in the plane of the hologram between the vibrations emitted by the object and the vibrations emitted by the coherent background. The variations in phase of the vibrations emitted by the object are transformed into variations in intensity, and it is these which are recorded on the photographic plate. The continuous variations of intensity cannot be reproduced mechanically; one must replace them by a large number of small elements, each one of these having a determinate position and dimensions. After development of the photographic plate, the binary hologram then behaves like an opaque screen pierced by a large number of apertures of determinate positions and dimensions. It is natural to assume that in a small region of the hologram the amplitude ought to be proportional to the dimensions of the local aperture. As to the phase, we shall see that it is determined by the position of the aperture.

Let us compare the binary hologram with a grating, and let us first consider phenomena in one dimension. Figure 8.13 represents a grating R formed by equidistant slits. The grating R is illuminated by a parallel bundle of light originating in a point source S at infinity. All the rays arrive in phase at the direct image F located at the focal

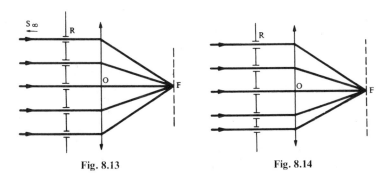

Fig. 8.13   Fig. 8.14

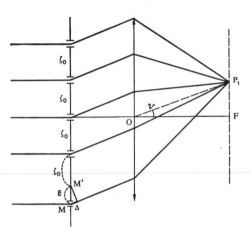

Fig. 8.15

point of the objective O. Nothing is changed if the slits are no longer equidistant (Fig. 8.14): all the rays still arrive in phase at F. This is no longer the case if one considers what happens in the neighborhood of a spectrum, say in the neighborhood of the first order spectrum P_1 (Fig. 8.15). If ζ_0 is the step of the slits which form the grating, then in the direction v corresponding to the first order spectrum one has (Section 2.6 and Chapter 4):

$$v\zeta_0 = \lambda \qquad (8.18)$$

Let M be a slit displaced by an amount ε with respect to the position M' pertaining to the grating. The ray diffracted by M and arriving at P then displays a path difference with respect to the others amounting to $\Delta = v\varepsilon$, from which, in accordance with (8.18), we have

$$\Delta = \varepsilon\frac{\lambda}{\zeta_0} \qquad (8.19)$$

Any displacement ε of a slit from its theoretical position produces a path difference $\varepsilon\lambda/\zeta_0$ and a phase difference $\varphi = 2\pi\varepsilon/\zeta_0$. This is the principle which is used to record the phase variations in a binary hologram.

Consider a general object: the letter A, for example (Fig. 8.16). One digitalizes the object, that is to say, one replaces the continuous lines of the letter A by a number of points sufficiently close together

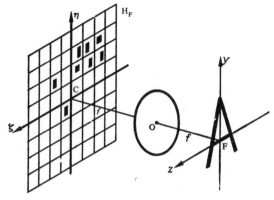

Fig. 8.16

to give the eye an impression of continuity. The computer calculates the Fourier transform of the object which is located in the plane $C\eta\zeta$. This transform itself is digitalized. One calculates the transform at a number of points at least equal to the number of points on the object. One then divides the plane $C\eta\zeta$ into a number of small square cells, a number equal to that of the points at which one has calculated the transform of the object. Each cell contains a small black rectangle on a white background. The dimensions of the rectangle determine the amplitude at the point considered and the phase is obtained by shifting the little rectangle with respect to the center of the cell. The whole assembly of small black rectangles is traced on a sheet of white paper by a printer controlled by the computer. One may replace the printer by a cathode-ray tube on whose screen the drawing appears. This is photographed, and after development the binary hologram behaves like an opaque screen pierced by a large number of small rectangular apertures occupying the place of the small black rectangles. The hologram thus constructed is illuminated normally by a parallel bundle (Fig. 8.17). In the direction of the incident rays, that is to say at F, the shifting of the apertures produces no effect and one observes at F an image of the point source. This is no longer the case in the direction of the first order spectrum $v = \lambda/\zeta_0$ corresponding to the step ζ_0 of the cells. The spectrum of the first order thus reconstructs an image of the object as well as the symmetric spectrum.

Because of its binary structure, the hologram will also reconstruct other images corresponding to spectra of higher order, such as are indicated by dashed lines in Fig. 8.17. One may note that the binary

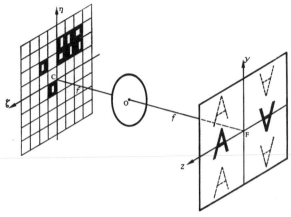

Fig. 8.17

character of the hologram eliminates all the difficulties arising from the question of the linearity of the emulsions.

8.13   Holograms at Several Levels. Kinoforms

When the computer has calculated the complex amplitude produced by the object in the plane of the hologram, it may add an auxiliary complex amplitude which plays the role of a coherent bundle. The computer calculates the resultant intensity, whose variations may be reproduced by a printer or a cathode-ray tube. The representation will the more closely approach the structure of a hologram as the number of levels of the printer or the cathode-ray tube is larger. The photograph of the drawing thus obtained gives the hologram.

In all holograms, one reconstructs two images of the object: one real and the other virtual. The holograms called "kinoforms" reconstruct only a single image of the object, and consequently are of some interest because all the flux is concentrated in this image.

We have spoken earlier of phase holograms (Section 8.9). Suppose that, using the computer, one constructs a drawing of variable blackening such that after bleaching the profile of the emulsion will be that of a Fresnel lens. This is a plane lens in which the variations in thickness are transformed into variations of phase which are always bounded between 0 and 2π. Such a lens may be considered as the hologram of a point object. On illuminating this lens by a bundle of parallel rays, one

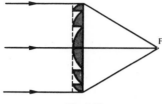

Fig. 8.18

evidently obtains a single image of the source at F. One may generalize this to the case of a general object; the hologram obtained, called a kinoform, behaves like an assembly of Fresnel lenses and reconstructs only a single image, whether real or virtual, of the object.

CHAPTER 9

Interferometry

9.1 Michelson Interferometer

In Chapter 2 we gave the fundamental laws of interference without being specific about the experimental layout of the interference apparatus. Figure 9.1 gives the principle of one of the simplest and most elegant types of apparatus: the Michelson interferometer. It consists of a semireflecting plate G and two plane mirrors M_1 and M_2. We shall, to begin with, neglect the thickness of G. The two mirrors M_1 and M_2 are mutually perpendicular and the plate G is inclined at 45° with respect to the normals to M_1 and to M_2, In Fig. 9.1, we have shown the traces of the elements M_1, M_2, and G, which are supposed perpendicular to the plane of the figure. The plate G, which is called the beam splitter, reflects as much light as it transmits. It plays the role of the wave-divider, of which we have spoken in Chapter 2. The interferometer is illuminated with monochromatic light by means of a bundle of rays parallel to the plane of the figure and normal to the mirror M_2. Let us follow the path of one of these rays through the apparatus. The ray SI is split up into two rays when it arrives at G and we have the following two paths:

(a) path 1: IA_1IT
(b) path 2: IA_2IT

These two rays, which have traversed different paths, are superimposed on leaving I and thus interfere. Consider the reflection M'_1 of M_1 in G.

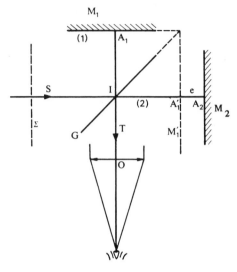

Fig. 9.1

Note that M'_1 and M_1 are symmetric with respect to G. Everything happens as if the incident ray SI had been reflected at M'_1 and M_2, M'_1 being a fictitious mirror. If e is the distance separating M'_1 and M_2, the path difference is $\Delta = 2e$ since M'_1 and M_2 are parallel. In order to observe the phenomena, place the eye at the focal point of an objective O which has been focused on M_1 or M_2. As a natural consequence of the interference of the two rays which overlap along IT, the intensity I along the direction IT is given by the fundamental laws of interference. If the mirrors M_1 and M_2 have the same reflection factor, the amplitudes of the interfering vibrations are equal and one has from (2.8):

$$I = I_0 \cos^2 \frac{\pi\Delta}{\lambda} = I_0 \cos^2 \frac{2\pi e}{\lambda} \qquad (9.1)$$

This formula is valid for any ray whatever of the bundle, since all the rays have the same direction, and the eye sees a uniform field. When one changes the distance e which separates M'_1 and M_2, one varies the intensity, the field always remaining uniform.

Instead of considering the rays, one may consider the waves. After traversing the interferometer, the incident plane wave Σ gives rise to two plane waves which are reflected at M_1 and $\dot{M}_2$. The two plane waves, which overlap in the plane of the receiver (here the retina), have a path difference which is constant and equal to $2e$.

9.2   Observation of the Fringes of Equal Thickness

Rotate the mirror M_2 in Fig. 9.2 very slightly around an axis O situated in its plane and perpendicular to the plane of the figure. The mirror M_2 makes an angle ε with M'_1. In this case the two wave-fronts Σ_1 and Σ_2 make an angle 2ε with each other. In the experiments, the distance between M'_1 and M_2 is always very small, and one may equally well consider that the eyepiece is focused on M_2 or on M'_1 (or M_1). At a general point A_2 located at a distance $x = OA_2$ from O, the path difference is $\Delta = 2\varepsilon x$, from which the intensity is:

$$I = I_0 \cos^2 \frac{2\pi\varepsilon x}{\lambda} \tag{9.2}$$

One sees straightline fringes, equidistant and parallel to the axis O. The bright fringes are given by

$$\frac{\Delta}{\lambda} = \frac{2\varepsilon x}{\lambda} = p \tag{9.3}$$

The distance separating two consecutive bright fringes is $\lambda/2\varepsilon$. For the black fringes, one has $\Delta/\lambda = p + \frac{1}{2}$ and the distance which separates two consecutive black fringes is also $\lambda/2\varepsilon$. Along a line $\Delta = $ constant the

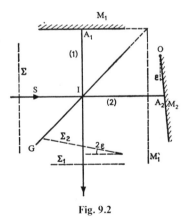

Fig. 9.2

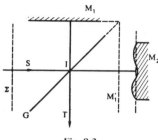

Fig. 9.3

intensity remains constant, and consequently the fringes show the lines of equal thickness of the air wedge enclosed between M_2 and M'_1.

Since a point source is used, the fringes are not localized; and one may, moreover, see them as if they were on M_2 and M'_1 (or M_1) by changing the focus.

Suppose that the mirror M_2 were deformed (Fig. 9.3) and let it have a mean position parallel to M'_1 ($\varepsilon = 0$). As a result of the deformations of M_2, the path difference $\Delta = 2e$ is no longer constant. The variations of intensity given by $\cos^2(2\pi e/\lambda)$ now characterize the deformations of M_2. The interference fringes thus enable us to determine with precision the deformation of M_2.

In the Michelson interferometer, the two paths of the bundles which interfere are well separated. It is therefore simple to introduce a transparent object in one of the light paths without affecting the other. The interference phenomena observed between the wave deformed by the object and the plane wave corresponding to the other light path permit one to study the transparent object.

9.3   Fringes of Equal Inclination

Let us return to the arrangement of Fig. 9.1 where M'_1 and M_2 are parallel. Illuminate the interferometer with rays having different inclinations and originating from a point source S_1 (Fig. 9.4). A general ray S_1I_1 gives rise to two parallel rays which are superimposed at a point P in the focal plane F of the objective O. It is for this reason that we are now going to observe these phenomena in the focal plane of the objective O. The intensity at P due to the interference of these two rays is equal to $\cos^2(\pi\Delta/\lambda)$, where Δ is the path difference at P. This path difference may readily be calculated as shown in Fig. 9.5, which

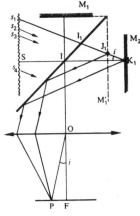

Fig. 9.4

reproduces a part of Fig. 9.4. From J_1, drop the perpendicular $J_1 H$ on to the ray reflected by M_2. One then has:

$$\Delta = \overline{J_1 K_1} + \overline{K_1 H} = 2e \cos i \qquad (9.4)$$

since, according to Malus's theorem, the optical paths traversed from H to P and from J_1 to P are equal. If $e = $ constant, the path difference depends only on the inclination i of the rays: the interference fringes are rings centered at F, the point of intersection of the normal to M_1 with the plane of the objective. Different sources such as S_2, S_3, S_4, etc. give rise to exactly the same phenomena at P. The sources $S_1, S_2, S_3, \ldots$ may be part of an extended incoherent source. In this case, all the interference effects to which they give rise are incoherent, but since all

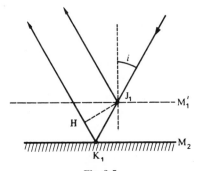

Fig. 9.5

these phenomena are identical, they can be superimposed without any difficulty. The interferometer may be illuminated by an extended mono-chromatic source, and the fringes are localized in the focal plane of the objective O. The order of interference at F is $2e/\lambda$ and at P it is $2e(\cos i)\lambda$. On passing from F to P the order of interference diminishes by the amount

$$\frac{2e}{\lambda} - \frac{2e \cos i}{\lambda} \simeq \frac{ei^2}{\lambda} \qquad (9.5)$$

Suppose that there is a bright spot at F. If, in going from F to P, one counts N bright rings, the angular radius i of the Nth bright ring is given by $(ei^2/\lambda) = N$, from which:

$$i = \sqrt{\frac{\lambda}{e}} \sqrt{N} \qquad (9.6)$$

The same formula is valid for the dark rings.

9.4   Visibility of the Fringes of Equal Thickness

Consider the arrangement shown in Fig. 9.2 and observe the fringes of a wedge of air ε. At a point situated at a distance x from O, the intensity for normal incidence is given by formula (9.2). If one illuminates the interferometer with a bundle of parallel rays making an angle i with respect to the primitive direction SA_2, the path difference is $2e \cos i = 2\varepsilon x \cos i$. It varies with the inclination of the rays.

Consider an extended monochromatic source placed at the focal point of the objective of a collimator. It sends an infinity of bundles of parallel rays into the interferometer. Therefore, in order that the pheno-mena shall remain sharp, it is necessary that at a general point of M_2 the path difference shall vary as little as possible with the inclination of the rays. For normal incidence, the path difference is $2e$ and for incidence at an angle i, it is $2e \cos i \simeq 2e(1 - \frac{1}{2}i^2)$. When one goes from normal incidence to incidence at an angle i, the path difference varies by the amount ei^2. Consequently, if e increases i ought to decrease. The greater the thickness of the air separating M'_1 and M_2, the smaller the source ought to be. This is a general result, regardless of the type of interferometer or the shape of the layer of air.

9.5 Temporal Coherence

The interferometer is now illuminated by a point source of non-monochromatic light (Fig. 9.6). One may equally well observe the fringes of equal thickness or the fringes of equal inclination. For simplicity, we consider the phenomena along one ray. An incident train of waves is divided into two wavetrains which follow the paths (1) and (2). If the path difference is longer than the length of the train of waves (coherence length), then on leaving the interferometer, the two wavetrains which originated in the same initial wavetrain will not overlap. There may, nevertheless, be other wavetrains which do overlap, but they will then have originated in two different initial wavetrains. Since the emission of wavetrains by the atoms is random, the phase difference between the wavetrains which overlap is also random.

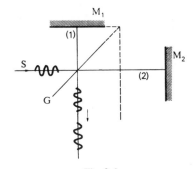

Fig. 9.6

Interference effects are no longer observable, and there is *temporal incoherence*. In order to be able to observe interference effects, it is necessary that the path difference be less than the coherence length. As the path difference becomes smaller compared to the coherence length, the fringes become so much the sharper. If the path difference is close to zero, for example if it is less than 1 μm, the fringes are visible even with very short wavetrains: the fringes are visible in white light.

9.6 White-Light Fringes

Up till now we have not taken into account the actual thickness of the beam splitter. The semireflecting surface corresponds to the dark line (Fig. 9.7). The rays that follow path (1) traverse G once and the

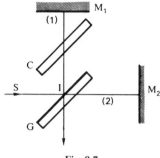

Fig. 9.7

rays that follow path (2) traverse G three times. In monochromatic light, one could compensate for this path difference, and attain at a zero path difference by increasing the separation between M_1 and G. But this can only be accomplished for one wavelength, because the index of refraction of the plate G (a plate of glass) varies with the wavelength. In order to make the two paths identical, regardless of the wavelength, one introduces a plate C, called the "compensator" into path (1). The plate C should have the same thickness and the same index as G. One may then obtain as small a path difference as one wishes, regardless of the wavelength, and observe the interference phenomena in white light. Everything that was said in paragraph 2.8 remains valid. If one tilts M_2 through a very small angle, one sees on M_2 (or M_1) colored fringes, and in those regions where the path difference exceeds several wavelengths, one sees a white of higher order.

9.7   Fringes Produced by Transparent Plates

The phenomena just discussed may be observed with glass plates or with layers of air. Consider a layer of air contained between two glass surfaces M_1 and M_2 (Fig. 9.8). The layer is illuminated by a parallel bundle of monochromatic light. An incident ray SI is reflected at A_1 on the surface M_1 and at A_2 on the surface M_2. We shall neglect the effects produced by the other faces of the layers (shown in dashed lines), for example, because they have been treated and do not reflect light (see Section 9.8). If one takes the intensity of the incident bundle as unity, the intensity reflected by M_1 has an intensity of the order 0.04 and the transmitted bundle an intensity equal to $1 - 0.04 = 0.96$. The bundle reflected by M_2 then has an intensity equal to 0.96×0.04,

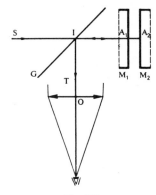

Fig. 9.8

which is again practically equal to 0.04. The bundles reflected three times have a negligible intensity. Thus the only bundles that matter are those reflected once by M_1 and M_2. The only difference between the arrangements shown in Figs. 9.1 and 9.8 is the replacement of the fictitious mirror M'_1 of Fig. 9.1 by a real reflecting surface located at M_1 in Fig. 9.8. One may therefore observe all the phenomena which have been described earlier, fringes of equal thickness or fringes of equal inclination. If the two plane surfaces M_1 and M_2 make a small angle, one observes the fringes due to an air wedge. In the case where M_1 is a spherical surface and M_2 a plane surface, the fringes of equal thickness are rings which are called Newton's rings. By measuring the diameters of the rings, one may deduce the radius of curvature of the spherical surface.

Now bring the two surfaces M_1 and M_2 into coincidence. In formula (9.1) it is necessary to set $e = 0$; however, instead of finding a maximum of light, one finds a minimum which in fact is zero. This occurs because the reflections at M_1 and M_2 are not of the same sort; at M_1 there is a glass–air reflection, while at M_2 there is an air–glass reflection. When the reflections are physically different as we have just said, it is necessary to add to the geometric path difference an additional path difference equal to $\lambda/2$ and one has:

$$\Delta = 2e + \frac{\lambda}{2}, \qquad I = I_0 \sin^2 \frac{2\pi e}{\lambda} \tag{9.7}$$

If one of the surfaces is plane and the other is deformed, one may measure the defects of the deformed surface with great precision. This

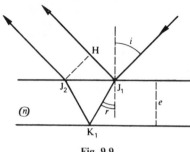

Fig. 9.9

is the principle of the interferometric control of the manufacture of plane surfaces.

One may observe the same phenomena with a plate of glass by causing two bundles reflected from the two faces to interfere (control of the parallelism of the two faces of a plate). If the two faces are plane and parallel, one observes the rings of equal inclination, but formula (9.4) is replaced by the following formula which may readily be derived (Fig. 9.9):

$$\Delta = 2n\overline{J_1K} - \overline{J_1H} = 2ne \cos r + \frac{\lambda}{2} \qquad (9.8)$$

where n is the index of the plate and r is the angle of refraction.

9.8   Diminution or Augmentation of the Reflecting Power by Means of Thin Transparent Films

One very well-known application of the interference phenomena produced by thin transparent films is shown in Fig. 9.10. By depositing a thin transparent film of index n and thickness e on a glass support of index N, one may suppress, or at least strongly diminish, the light

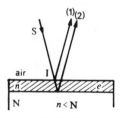

Fig. 9.10

reflected by the glass of index N. By depositing thin layers of this type on all the glass surfaces of an optical instrument, one can increase the transparency of the instrument at the same time as one diminishes the parasitic light caused by the multiple reflections of the light reflected by the glass surfaces.

As a result of reflections at the two interfaces, air–thin layer and thin layer–glass, an incident ray gives rise to two reflected rays (1) and (2). The path difference is $\Delta = 2ne$ if $n < N$. Let a and b be the amplitudes of the rays (1) and (2). The intensity of the reflected light is given by formula (2.10). There is a minimum of reflected light if $\varphi = 2\pi\Delta/\lambda = \pi$, that is if $ne = \lambda/4$. In order that this minimum be in fact zero, it is necessary that $a = b$, which is not attainable in practice with the indices of substances known at the present time. One can nevertheless bring about a reduction by a factor of 20 in the intensity of the light reflected by a glass surface.

It is readily seen that if, on the other hand, $n > N$ and $ne = \lambda/4$, one in fact increases the light reflected by the glass surface. In fact, one has $\Delta = 2ne + \frac{1}{2}\lambda$ since $n > N$ and consequently $\Delta = \lambda$. The two reflected rays are in phase, and there is a maximum of reflected light.

Thin films are deposited by evaporation in a vacuum. They are of great interest from a fundamental point of view and also because of their many applications, notably in electronics (microminiaturization).

9.9   Multiple-Beam Interferometers: Fabry–Perot Interferometer

In the preceding paragraph, we did not take into account multiple reflections, because the bundles reflected more than once were of negligible intensity. If the surfaces at which the light is reflected are semireflecting, it is necessary to take into account multiple reflections. Consider two identical parallel semireflecting surfaces M_1 and M_2 (Fig. 9.11). The reflection factor R and the transmission factor T of the surfaces are defined by the following ratios:

$$R = \frac{\text{reflected energy}}{\text{incident energy}} \qquad (9.9)$$

$$T = \frac{\text{transmitted energy}}{\text{incident energy}} \qquad (9.10)$$

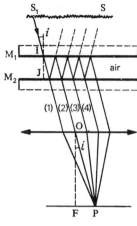

Fig. 9.11

We shall consider here the reflection factor of *the air with respect to the
surface* M_1 (or M_2) and shall take the intensity of the ray IJ as unity.
Since the media are transparent, all the incident energy reappears in
the reflected energy and the transmitted energy:

$$R + T = 1 \qquad (9.11)$$

If reflection takes place at a semireflecting metallic layer, a part A
of the energy is lost by absorption and we have:

$$R + T + A = 1 \qquad (9.12)$$

An incident ray S_1I coming from a point source S_1 gives rise to transmitted rays (1), (2). (3), etc. These rays are parallel, and they are
combined at a point P in the focal plane of the objective O, where they
interfere. If ε is the path difference corresponding to the reflection at
the air–surface M_1 (or M_2) interface, i the angle of incidence and e the
thickness of the layer of air separating M_1 and M_2, the path difference
between two successive rays is, in accordance with (9.4), equal to:

$$\Delta = 2e \cos i + 2\varepsilon \qquad (9.13)$$

The semireflecting surfaces M_1 and M_2 belong to two glass plates with
parallel faces. We shall designate as T the transmission factor of either
of these plates, which are supposed identical. This transmission factor
T evidently takes into account the presence of the semireflecting
surface. It does not depend on the direction of travel of the light. It is

not necessary to take into account the change of phase in the course of transmission produced by the two plates, because it is the same for all the rays.

The *complex amplitudes* of the transmitted rays are:

(1) T,   (2) $TRe^{-j\varphi}$,   (3) $TR^2e^{-2j\varphi}$, $\ldots$

where $\varphi = 2\pi\Delta/\lambda$, and the total amplitude at P is:

$$U_P = T(1 + Re^{-j\varphi} + R^2e^{-2j\varphi} + \cdots) = \frac{T}{1 - Re^{-j\varphi}} \qquad (9.14)$$

The intensity at P is given by

$$I = U_P U_P^* = \frac{T^2}{1 - 2R\cos\varphi + R^2} \qquad (9.15)$$

Using the identity:

$$1 + R^2 - 2R\cos\varphi = (1 - R)^2 + 4R\sin^2\frac{\varphi}{2} \qquad (9.16)$$

one has:

$$I = \frac{T^2}{(1 - R)^2} \frac{1}{1 + [4R/(1 - R)^2]\sin^2(\varphi/2)} \qquad (9.17)$$

The phase difference φ and the intensity I depend only on the inclination i of the rays; the fringes are rings centered on the intersection of the normals to the plates with the focal plane of the objective O. One may reason as in paragraph 9.3: the fringes, localized in the focal plane of the objective O, are perfectly sharp with an extended incoherent source. Figure 9.12 shows the structure of the fringes: if R is large the

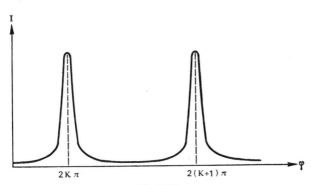

Fig. 9.12

maxima corresponding to $\sin^2(\varphi/2) = 0$ appear in the form of very sharp bright rings standing out against an almost black background. Comparing formulas (9.4) and (9.13), one sees that the formula giving the radii of these rings is the same as before, formula (9.6).

One may also observe the multiple-beam fringes due to reflection, but the calculations are much more complicated because the first ray reflected by M_1 has not traversed the semireflecting surface M_1 and does not belong to the same series as the others which have traversed it twice.

9.10   Interference Filters

Form a layer of air of very small thickness e, for example $e = 0.54 \, \mu m$ and illuminate it with white light. In the direction perpendicular to the layer one has a maximum for wavelengths such that $\sin^2(\varphi/2) = 0$. According to (9.13) and taking $\varepsilon = 0$, these wavelengths for normal incidence are:

$$\lambda_1 = 2e = 1.08 \, \mu m, \quad \lambda_2 = \frac{2e}{2} = 0.54 \, \mu m, \quad \lambda_3 = \frac{2e}{3} = 0.36 \, \mu m, \ldots$$

If R is sufficiently large, the interferometer passes only the wavelength $\lambda = 0.54 \, \mu m$. The other maxima correspond to wavelengths outside of the visible spectrum. Besides this, the path difference $\Delta = 2e \cos i$ varies very little with the angle of incidence i, since e is very small. This means that the wavelength of $0.54 \, \mu m$ (green light) transmitted by the interferometer is practically independent of the inclination of the rays. Looking through the interferometer in different directions, the color does not change. In practice, the layer of air is replaced by a thin transparent solid layer having semireflecting faces. The assembly constitutes a transmission type interference filter. One may choose the transmitted wavelength in the visible spectrum or outside the visible. One obtains a heat reducing filter if the heat radiations are not transmitted by the filter. One may equally well realize reflection type interference filters using the same principle. A "cold mirror" is one which reflects the visible but not the infrared.

Reflection type interference filters may be obtained by photographing stationary waves (Section 2.9). Illuminate a very fine grained sensitive layer (photographic film) (Fig. 9.13) with a bundle of parallel rays (monochromatic light). In all the antinodal planes, the silver is

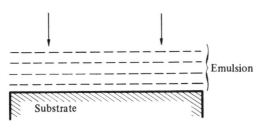

{ Emulsion

Substrate

Fig. 9.13

reduced and one obtains a series of semireflecting lamellae of reduced silver, spaced at equal distances of $\lambda/2$. Illuminate it with white light: for radiation of wavelength λ, the vibrations reflected by all the planes are in phase, since on reflection the path difference between two ventral planes is twice $\lambda/2$. For another radiation, even one that is very close to the first, this is no longer true if there is a large number of lamellae. The phase differences in fact take on a large number of values between 0 and 2π, and the vibrations undergo destructive interference. The photographic plate, when illuminated by white light, in practice reflects only the light of the wave length that was used to expose it. The process invented by Lippman in 1891 for color photography was based on this principle.

The colors of hummingbirds (Trochilidae) and of certain butterflies (Morpho) are also produced by means of superposed lamellae and are not due to pigments.

9.11 Interference Spectroscopy

When a Fabry–Perot interferometer is illuminated by a source emitting two wavelengths λ and λ', one obtains two systems of rings. If the plates of the interferometer have a large reflecting power, the rings are very sharp and one can separate them. This is the principle of the interference spectroscope based on the use of the Fabry–Perot interferometer.

One may record all the elements of a spectrum simultaneously on a single photographic plate, or, on the other hand, we may collect these elements in succession. Figures 9.14 and 9.15 refer to the first method. The source whose spectrum one wishes to study illuminates the Fabry–Perot interferometer (F.P. in Fig. 9.14), and by means of an objective O one projects the rings on the slit of a spectroscope. The slit F is in the

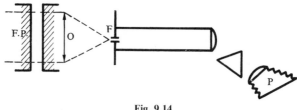

Fig. 9.14

focal plane of the objective and is of a suitable size. One obtains on the photographic plate P as many images of the slit F as there are different wavelengths. Each image gives a portion of the rings which correspond to a particular wavelength (Fig. 9.15). Due to the sharpness of the rings, one can display the structure of the lines. Here the spectrograph only serves to separate the lines from another, and it is the interferometer which permits one to reveal in each image of the slit the fine structure of the line being studied.

In the second procedure, the spectral line whose structure one wants to determine is first isolated by means of a monochromator. The emerging bundle is allowed to fall on a Fabry–Perot interferometer and one observes the rings in the focal plane of an objective. An annular diaphragm concentric with the rings is placed in the focal plane and isolates a small portion of the rings. Behind the diaphragm is placed a photomultiplier which is connected to an amplifier and to a read-out instrument. On changing the spacing of the plates the rings expand or diminish, and they march past the annular diaphragm. The variations of light intensity are recorded by means of the photomultiplier, and one obtains a curve which gives the structure of the line being studied. In practice, the variation in the path difference is not obtained by changing

Fig. 9.15

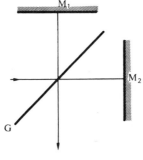

Fig. 9.16

the spacing of the plates, but by causing a variation in the pressure in an enclosure in which the interferometer is placed.

Another important interference method is Fourier transform spectroscopy. In this method, the source to be studied illuminates a two-beam interferometer, for example a Michelson interferometer (Fig. 9.16). Let us characterize the spectral composition of the source by a function $B(\sigma)$ which gives the variation of the intensity as a function of the wave number $\sigma = 1/\lambda$. If the interferometer is illuminated by a source emitting in a very small interval $d\sigma$, then according to (9.1) the intensity at the exit of the interferometer is:

$$B(\sigma) \cos^2(\pi \, \Delta\sigma) \, d\sigma = \tfrac{1}{2}B(\sigma)(1 + \cos 2\pi \, \Delta\sigma) \, d\sigma \qquad (9.18)$$

The variable part is, to within a constant factor:

$$I(\Delta) = B(\sigma) \cos(2\pi \, \Delta\sigma) \, d\sigma \qquad (9.19)$$

If the interferometer is illuminated by a source having an arbitrary spectral distribution, one has:

$$I(\Delta) = \int_0^\infty B(\sigma) \cos(2\pi \, \Delta\sigma) \, d\sigma \qquad (9.20)$$

$I(\Delta)$ is called the interferogram. If the source emits a monochromatic radiation, the interferogram is a sinusoid of constant amplitude. In the case of nonmonochromatic light, the amplitude of the sinusoid decreases when Δ increases, which is in agreement with the preceding results (Section 9.5). In order to determine $I(\Delta)$ one causes the path difference Δ to vary progressively; one records the intensity variations of the emerging bundle, that is to say $I(\Delta)$ plus a constant, by means of a Golay cell for example. Consider the curve $B(-\sigma)$ symmetric to

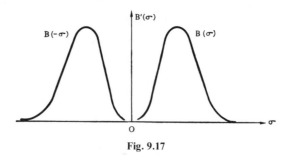

Fig. 9.17

$B(\sigma)$ (Fig. 9.17). Obviously, negative wave numbers have no physical reality whatever. One may write:

$$B'(\sigma) = \tfrac{1}{2}[B(\sigma) + B(-\sigma)] \tag{9.21}$$

from which:

$$I(\Delta) = \int_{-\infty}^{\infty} B'(\sigma) \cos(2\pi \, \Delta\sigma) \, d\sigma \tag{9.22}$$

The Fourier cosine transform then permits us to write:

$$B'(\sigma) = \int_{-\infty}^{\infty} I(\Delta) \cos(2\pi \, \Delta\sigma) \, d\Delta \tag{9.23}$$

One can, therefore, calculate the spectrum emitted by the source as a function of the interferogram.

9.12   Holographic Interferometry

Let us consider a transparent object A, for example, a glass plate of thickness e and index n which has some irregularities of thickness (Fig. 9.18). It is illuminated by a parallel bundle, and the photographic plate H also receives the coherent plane reference wave Σ_R. Let $F_1(\eta, \zeta)$ be the amplitude produced at a general point η, ζ of H by the wave Σ_1 which has traversed the object A. At the same point, the coherent wave Σ_R produces the amplitude $ae^{-jK\theta\zeta}$. Make an exposure under these conditions. The plate H receives the intensity:

$$\begin{aligned}
I_1 &= (ae^{-jK\theta\zeta} + F_1)(ae^{jK\theta\zeta} + F_1^*) \\
&= a^2 + |F_1|^2 + ae^{jK\theta\zeta}F_1 + ae^{-jK\theta\zeta}F_1^*
\end{aligned} \tag{9.24}$$

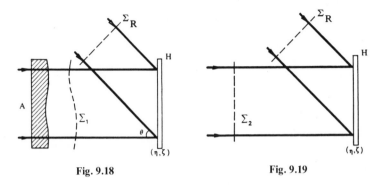

Fig. 9.18   Fig. 9.19

Make a second exposure with the same exposure time, but without the object A (Fig. 9.19). The plate H receives the coherent plane wave Σ_R and the plane wave Σ_2 which corresponds to the other bundle and produces the amplitude $F_2(\eta, \zeta)$. The intensity is:

$$I_2 = (ae^{-jK\theta\zeta} + F_2)(ae^{jK\theta\zeta} + F_2^*)$$
$$= a^2 + |F_2|^2 + ae^{jK\theta\zeta}F_2 + ae^{-jK\theta\zeta}F_2^* \qquad (9.25)$$

In toto, the plate receives the intensity $I_1 + I_2$. Develop the hologram and illuminate it with the wave Σ_R which was employed during the recording. From Section 8.6 the amplitude transmitted by the hologram is:

$$tae^{-jK\theta\zeta} = [t_0 - \beta'(F_1^2 + F_2^2)]ae^{-jK\theta} - \beta'a^2(F_1 + F_2)$$
$$- \beta'a^2e^{-jK2\theta\zeta}(F_1^* + F_2^*) \qquad (9.26)$$

The hologram reconstructs the virtual image $\beta'a^2(F_1 + F_2)$, that is to say, the two waves Σ_1 and Σ_2 corresponding to F_1 and F_2 (Fig. 9.20). One may set:

$$F_2 = b, \qquad F_1 = be^{-j\varphi} \qquad (9.27)$$

where φ represents the variations of phase produced by the plate A:

$$\varphi = \frac{2\pi}{\lambda}(n - 1)e \qquad (9.28)$$

In the virtual image, one has the amplitude $b + be^{-j\varphi}$ from which the intensity is:

$$b^2(1 + e^{-j\varphi})(1 + e^{j\varphi}) = 4b^2 \cos^2 \frac{\varphi}{2} \qquad (9.29)$$

One observes fringes which trace out the lines $(n - 1)e = $ constant.

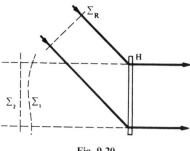

Fig. 9.20

The remarkable fact to which we call attention is the following: *the two waves* Σ_1 *and* Σ_2 *recorded at different times are nevertheless capable of interfering with each other.*

Another fundamental aspect of holographic interferometry is the possibility of studying diffusing objects, which is not possible as a practical matter in ordinary interferometry. One takes the hologram of the diffusing object and, after development, one repositions the hologram in exactly the same place, while keeping unchanged the experimental arrangement which had been used for the original recording. Through the hologram, one sees the real object and its virtual image as given by the hologram. These coincide exactly. If one slightly deforms the real object, the coincidence no longer exists. The interference between the real object and its virtual image brings about the appearance of fringes which characterize the deformation of the real object.

9.13   Speckle Interferometry

When laser light shines on a diffusing object, the object appears to be overlaid with a very fine granular structure. All the points of the object are coherent, and they send to the retina vibrations which are capable of interfering. Each point of the object forms on the retina an image which is the characteristic diffraction pattern of the optical system of the eye. It is the interference of these diffraction patterns which brings about the granular appearance called "speckle." The phenomenon is unchanged if one replaces the eye by a camera; after development, the image displays a speckle pattern whose spot size depends on the aperture of the objective. The greater the aperture, the finer the structure of the speckle; this is quite normal, since the size of the diffraction pattern of the objective decreases as its aperture increases. But it is not necessary to form an image of an object illuminated by a laser in order to obtain speckle. A diffusing object illuminated by a laser produces a speckle pattern in the whole region around it. In order to record a speckle pattern, one need only place a photographic plate at some distance from the diffusing object. By analogy with diffraction phenomena, one may say that the speckle is of the Fraunhofer type in the first case, and of the Fresnel type in the second case.

Consider now the first case. The diffusing object A (Fig. 9.21) is illuminated by a laser, and an image A' of A is formed on the photographic plate H by means of an objective O. For simplicity, we shall assume that the object A is a plane object. The granular structure of the image A', that is to say, the speckle, may be represented by a random function $D(\eta, \zeta)$ of the coordinates η, ζ of a point lying in the plane of A'. The function $D(\eta, \zeta)$ represents the variations of *intensity*.

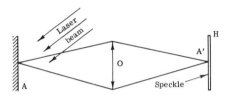

Fig. 9.21 Speckle in the image A' of a diffusing object A illuminated by a laser.

Now make two successive exposures with equal exposure times, but subject the photographic plate H to a small translation between the two exposures. For example, let the plate be moved by an amount ζ_0 in the direction of the ζ-axis. The intensity recorded on the plate is the sum of the intensities recorded during each exposure and may be written:

$$D(\eta, \zeta) + D(\eta, \zeta - \zeta_0) \tag{9.30}$$

Since a translation is equivalent to a convolution with a delta function, we may write the recorded intensity in the form:

$$D(\eta, \zeta) \otimes [\delta(\eta, \zeta) + \delta(\eta, \zeta - \zeta_0)] \tag{9.31}$$

where $\delta(\eta, \zeta)$ is a delta function centered on (η, ζ). As we have seen in Chapter 8 (holography), the *amplitude* transmitted by the negative after development may be written:

$$t = a - b\{D(\eta, \zeta) \otimes [\delta(\eta, \zeta) + \delta(\eta, \zeta - \zeta_0)]\} \tag{9.32}$$

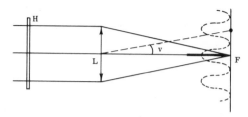

Fig. 9.22 Observation of the spectrum of the negative H on which have been recorded two identical speckles which are displaced with respect to one another.

where a and b are two constants characterizing the photographic emulsion that has been used. Let us now observe the spectrum of this negative, using the experimental setup shown in Fig. 9.22. The negative H is illuminated by a parallel bundle of light originating in a point source which emits monochromatic light of wavelength λ. One observes the spectrum of the negative H in the focal plane of the lens L. This spectrum is the Fourier transform of the amplitude t transmitted by H. This transform $\tilde{t}$ is:

$$\tilde{t}(u, v) = a\delta(u, v) - b\tilde{D}(u, v)(1 + e^{j(2\pi v\zeta_0/\lambda)}) \qquad (9.33)$$

Here u and v are the angular coordinates of a general point P in the focal plane of L, where one observes the phenomena. The first term $a\delta(u, v)$ of the right-hand side of (9.33) represents the image of the point source if one neglects diffraction, i.e. the image according to geometrical optics. This image is localized at the focal point F, and is very small. In the second term, aside from the constant b, the transform $\tilde{D}(u, v)$ of $D(\eta, \zeta)$ is modulated by the factor $1 + e^{j(2\pi v\zeta_0/\lambda)}$. Since the structure $D(\eta, \zeta)$ is very fine, its transform $\tilde{D}(u, v)$ is spread out over a large region of the focal plane of the lens L. If one neglects the image at F of the point source, then in all the rest of the focal plane the luminous intensity is given, to within a constant factor, by:

$$I = |\tilde{D}(u, v)|^2 |1 + e^{j(2\pi v\zeta_0/\lambda)}|^2 = |\tilde{D}(u, v)|^2 \cos^2\left(\frac{\pi v\zeta_0}{\lambda}\right) \qquad (9.34)$$

$|\tilde{D}(u, v)|^2$ has a very fine structure, like $D(\eta, \zeta)$. This diffused background $|\tilde{D}(u, v)|^2$ is modulated by $\cos^2(\pi v\zeta_0/\lambda)$, which represents Young's interference fringes. The angular distance separating two consecutive bright fringes is equal to λ/ζ_0. For example, with a translation ζ_0 equal to 20 μm the angular separation between two consecutive bright fringes is equal to $1°42'$, which is more than three times the apparent diameter of the sun.

Till now, it was the photographic plate H which was moved between the two exposures. The results would be exactly the same if the plate H were kept fixed, and the object A moved between the two exposures. If the translation of the object A between the two exposures were such that the displacement of the image A' were equal to ζ_0, the angular separation of the fringes in the spectrum of the negative would be λ/ζ_0. Suppose now that one part of the object A were displaced between the two exposures by an amount ζ'_0 different from the displacement ζ_0 of the rest of the object. Then, in the spectrum of the negative, the fringes produced by these two regions of the object will not be the same. From this, there arises the possibility of detecting, and measuring displacements or deformations of diffusing objects, for which the methods of classical interferometry are practically useless. This is the principle of speckle interferometry, of which we now proceed to give an example.

9.14   Measurement of Deformations or Displacements of Diffusing Objects by Speckle Interferometry

The principle of the experiment described in the preceding section clearly shows that we are concerned with studying deformations or displacements in a direction perpendicular to the axis of the optical system used for the observations. For example, in Figs. 9.21 and 9.22 the deformations or displacements of the object A must be in a direction perpendicular to the axis AA' of the objective O in order to be measurable by speckle interferometry. This is not meant to imply that it is impossible to study longitudinal deformations or displacements, which lie in a direction parallel to AA'. But these measurements pose more subtle problems which are beyond the scope of this book.

For simplicity, consider the object A as composed of two parts A_1 and A_2 (Fig. 9.23). The region A_1 is supposed fixed and the region A_2 is displaced between the two exposures. Let this displacement be a transverse displacement equal to ζ'. Make two successive exposures on the plate H, which is subjected to a translation ζ_0 between the two exposures. In the region A'_1, the plate H records two identical speckles displaced from each other by an amount ζ_0, since the region A_1 is fixed. On the other hand, in the region A'_2 the plate H records two speckles which are identical, but displaced from each other by a different

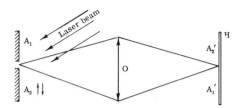

Fig. 9.23 Study of the transverse displacement of a diffusing object by speckle interferometry.

amount. In our example, the region A_2 is translated an amount ζ' and the plate H is given a parallel translation ζ_0 between the two exposures. This is a special case, but it does not reduce the generality of the method. After development, examine the negative, using the experimental arrangement of Fig. 9.24. The region A_1' gives rise to fringes

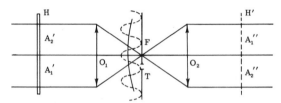

Fig. 9.24 Detection of displaced or deformed regions of the diffusing object.

which are shown by a solid line in the diagram, and the region A_2' gives rise to fringes which are shown by a dashed line. A second objective O_2 forms an image of H at H'. In the focal plane of the objective O_1 place an opaque diaphragm containing a slit T which coincides with a zero of the fringe pattern given by the region A_1'. In the image H' the image A_1'' disappears. On the other hand, the slit T in the diaphragm passes some of the light coming from the region A_2''. The image A_2'' is visible. Thus this method permits one to see those regions of the object A which are deformed or displaced, all the fixed regions being eliminated.

In order to make measurements, one may proceed as shown in Fig. 9.25. Instead of observing the image of the object itself, one examines the fringes in the spectrum of the negative in the plane of the lens L by placing a diaphragm containing a hole T in contact with the negative H. The orientation and the spacing of the fringes will vary with the position of the hole T, that is to say, they will vary with the

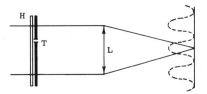

Fig. 9.25   Point-by-point study of the displacements or deformations of a diffusing object.

position of the point being examined in the object *A*. One may thus measure the direction and the magnitude of the displacement of different regions of the object point by point.

This very simple method has found many applications in different fields. One may measure the displacements or the deformations of real objects, such as, for example, the cement joints between pre-fabricated elements used in the construction of buildings. It has been applied to the measurement of the roughness of surfaces, to measurement of the vibration of diffusing objects, to image processing, etc. We must also call attention to the remarkable application of speckle interferometry to astronomy. In this problem, the coherent source is the star, and the diffuser producing the speckle is the atmospheric turbulence. If one records the image of a star with an exposure time sufficiently short to "freeze" the atmospheric turbulence, this latter will strongly perturb the form of the wave surface which enters the telescope, and the image of the star will be composed of a large number of small granules. These granules constitute an actual speckle, and their diameter is of the order of the size of the diffraction pattern corresponding to *the entire aperture* of the telescope. Starting from this speckle, one can study double stars, and in particular, one can study the apparent diameter of stars by a novel and particularly fruitful method.

9.15   Applications of Interference Methods

We have already spoken of the manufacture of thin films using interference phenomena, on the one hand to diminish, and on the other hand to augment, the reflecting power of glass surfaces. Interference spectroscopy is also a particularly important application (Section 9.11). Interference methods are also used for making extremely precise measurements. One uses them to determine the planarity or the quality of the polish of a glass or metal surface, the parallelism of plates, the radii of curvature of spherical surfaces, the form of a wave surface, the variations of pressure in an aerodynamic wind-tunnel, etc. Interference microscopes are among the instruments currently used in laboratories and in industry to study transparent or reflecting objects. Finally, in the domain of diffusing objects, the methods of speckle interferometry have given rise to numerous applications which are particularly simple and interesting.

Interference methods are well suited to making very precise measurements of length, and these are the methods on which were based the determination of the meter in terms of wavelengths. The first experimental determination of the length of the meter in terms of wavelengths was carried out by Michelson in 1892 at the Bureau International des Poids et Mesures with the interferometer which bears his name. In 1913, Benoit, Fabry and Pérot undertook anew the measurement of the standard meter using fringes of superposition. The red line of cadmium, utilized by Michelson, Benoit, Fabry and Pérot has a coherence length of less than a meter, which made the measurements difficult. A source emitting a line having a greater coherence length has replaced the red line of cadmium, and the Comité Internationale des Poids et Mesures has decided that the meter shall be defined thus: the meter is the length equal to 1,650,763.73 vacuum wavelengths of the radiation of the atom of Krypton 86 corresponding to the transition between the levels $2p_{10}$ and $5d_5$. Interference measurements using very long path differences are very easy with lasers, and it is probable that the definition of the standard meter will be modified so as to use a laser as a source.

CHAPTER 10

Elements of the Optical Processing of Information

10.1 Introduction

The experiments of Abbe on the filtering of spatial frequencies may be considered as one of the first experiments on the "optical processing of information." The information is an object illuminated by a point source (coherent illumination), and one studies the structure of the image as a function of the modifications which one causes its spectra to undergo. Abbe made use of simple screens to accomplish the filtering, and since the appearance of holography it has been possible to construct complex filters which are usable not only in coherent light, but also in incoherent light. Computers also have made possible important progress, because the purely experimental construction of filters is not always an easy, simple thing, while with digital calculations using the computer, the problem is generally soluble no matter what may be the type of filter considered. Finally, the transition from calculation to construction does not present too great difficulties.

Thanks to the development of filters, some new methods of processing of images have been brought into use. To begin with, we shall cite the method of form recognition which is a generalization of the method of signal detection used in communication theory. One can

145

detect the presence of a particular detail amongst all the other details constituting a luminous object. It is necessary to carry out a convolution operation between the signal to be detected (the luminous detail) and a filter which has been constructed based on the signal sought. The filter is a hologram, and in the plane conjugate to the object, one observes three images, of which one gives the solution to the problem.

In medicine and in biology for example, there are numerous applications to the recognition and to the separation of superimposed or adjacent chromosomes, to the automatic processing of radiography, etc. It is possible to carry out automatic reading of documents, which permits direct introduction into the computer of typewritten data, resulting in a considerable gain in time as compared with the current methods of introducing data into the computer.

Optics also plays a greater and greater role in the problems of information storage, thanks particularly to the properties of photographic emulsions, which are capable of storing 10^8 bits of information on 1 cm^2 of surface. The images may be recorded on the photographic plate by juxtaposition or by superposition. In this last case, one "takes out" the images separately by means of holographic filtering.

And finally, the methods of optical processing are applicable to electrical signals immersed in noise. They permit, for example, the processing of the return signal of a radar or even of a sonar.

The optical processing of information can thus have important practical consequences, and it is reasonable to expect notable improvements, especially if the connection between modern optics, information theory, electronics and computers becomes even more effective.

10.2   Filtering in Coherent Illumination

Figure 10.1 gives the basic arrangement for filtering in coherent light. The signal to be processed is recorded on a photographic film. It may, for example, be the sound track of a sound film, or a photograph of a text or of a landscape. We shall call the signal to be processed the "object" in order to retain the terminology of optics. It is in the plane A, and the coordinates of its points are referred to a system of axes η, ζ. The object A is illuminated by a parallel bundle (monochromatic light) and is located in the object focal plane of a lens O_1. A second lens O_2 gives an image of A at A'. For simplicity, we may suppose that

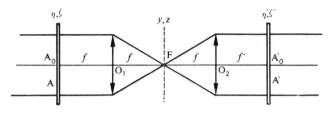

Fig. 10.1

the two lenses O_1 and O_2 are identical. The image focal point F of O_1 coincides with the object focal point of O_2, and the image A' is located in the image focal plane of O_2. Since the object A is at a distance from O_1 equal to the focal length of this lens, the Fourier transform of the object A is found in the image focal plane of O_1, that is to say, in the plane passing through F (Section 1.4). This Fourier transform, called the *spectrum of the object*, is nothing other than the diffraction pattern produced by the object A. It is in the plane passing through F that we are going to place the filter intended to modify the spectrum of the object, and consequently the structure of the image at A'. Let us specify the details of the object; we shall take, to begin with, a photograph of a periodic object having an intensity profile given by $1 + a\cos(2\pi\zeta/p)$, where $1/p$ is the spatial frequency and a is a positive quantity less than 1.

If we work in the straightline portion of the curve $t = f(W)$ (see Section 8.2) after development, the amplitude transmitted by the negative may also be represented by the expression $1 + a\cos(2\pi\zeta/p)$. As in certain types of photogravure, one replaces the continuous function $1 + a\cos(2\pi\zeta/p)$ by discrete values sufficiently closely spaced to continue to give an impression of continuity. One obtains this representation by multiplying $1 + a\cos(2\pi\zeta/p)$ by a Dirac comb, comb(ζ/ζ_0) of step $\zeta_0 \ll p$ (Fig. 10.2). The photograph, that is to say the object, is formed of very small dots of the same diameter which are more or less transparent (halftone photography). The spectrum of this object is given by the convolution of the transforms of $1 + a\cos(2\pi\zeta/p)$ and of comb(ζ/ζ_0). This spectrum is found in the plane passing through F (Fig. 10.1), and its structure is indicated in Fig. 10.3 (see Appendix A). Let us place at F a filter consisting of a simple slit of angular width v_0 parallel to the fringes of the photograph. Let us assume that v_0 is equal to $\lambda/2\zeta_0$. This screen passes only the central signal represented in Fig. 10.4. In order to find the image at A' (Fig. 10.1) it is necessary to take the Fourier transform of this signal. This is evidently an expression

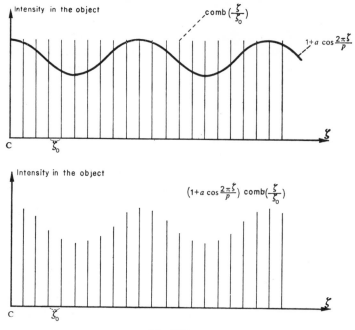

Fig. 10.2

of the form $1 + a\cos(2\pi\zeta'/p')$ which gives an intensity variation of the form $[1 + a\cos(2\pi\zeta'/p')]^2$. The discontinuous structure of the object no longer exists in the image. The filter we have used is a low-pass filter which stops the high spatial frequencies, and in particular the spatial frequency of the Dirac comb. The halftone structure has been erased. We have considered a very special case of an object containing

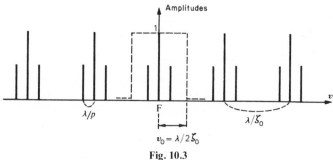

Fig. 10.3

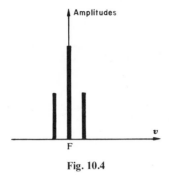

Fig. 10.4

only two spatial frequencies. In the case of a general object contaminated by noise, caused for example by granularity, a low-pass filter can reduce the granularity, but at the same time it eliminates the high spatial frequencies of the object, and this degrades the image. On the other hand, if a general object is contaminated with a periodic noise, one can suppress the noise by utilizing a filter which eliminates only that frequency or those frequencies which correspond to the periodic noise. The high frequencies are not entirely stopped and the image is less "degraded."

10.3   Form Recognition by Means of Autocorrelation

It is often necessary to determine what resemblance exists between two figures, or between one figure and a specific region of another figure. Problems of this sort are encountered in the identification of characters or of patterns, for example, when one wishes to compare two photographic maps of which one serves as a reference. The inverse problem of the detection of differences between two photographs of the same object taken at two different times is equally interesting.

To begin with, we shall set ourselves the following problem: given a collection of different signals, to recognize the presence of a particular signal. For example, one has at one's disposal a text which has a great number of characters, and the problem is to recognize the presence of a specific letter or word. The text to be examined is given in the form of a photograph with transparent letters on a black background. The experimental arrangement is that of Fig. 10.1, and the photograph of the text is at *A*. The recognition of the letter or the word is carried out by means of a filter consisting of a Fourier hologram of the letter

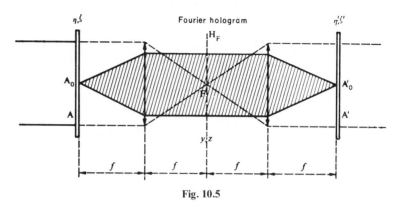

Fig. 10.5

or word to be identified. This hologram is placed at F in the experimental setup (Fig. 10.5). The identification takes place in the image plane A' of the object A.

Let us take a concrete example: one wishes to recognize the presence of the character 术 in a Chinese text. First construct the Fourier hologram $f(y, z)$ of this character. We take a photograph in such a way as to have on the negative a transparent character on a black background. The hologram of this negative is taken according to the schematic outline of Fig. 8.10 by placing the negative at A in this figure. After this, place the Fourier hologram H_F of the character 术 at the point F in Fig. 10.5 and examine the plane $\eta'\zeta'$. In order to understand the structure of the experiment, we are going to work out two cases:

(1)   The object A consists of a character 术 which is the same as the character which was used to form the Fourier hologram H_F.

(2)   The object A consists of the character 术 plus other different characters.

To begin with, consider the first case. We have the arrangement shown in Figs. 8.11 and 8.12. The Fourier hologram H_F is illuminated by a parallel bundle, and it reconstructs three images in the focal plane η', ζ' of the objective O_2. From (8.17) one has (Fig. 10.6):

(a)   a central luminous spot at A'_0,

(b)   an image (1) of the signal 术 corresponding to the wave $ae^{jK\theta z}f$, and

(c)   an image (2) of the signal 术 corresponding to the wave $ae^{-jK\theta z}f*$.

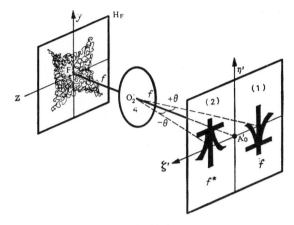

Fig. 10.6

Just in front of the Fourier hologram, the amplitude is given by the Fourier transform of the character 木. If one employs the same experimental conditions as at the time of recording the Fourier hologram of the character 木, the transform of this character is given by $f(y, z)$. The amplitude transmitted by the hologram is equal to the product of the incident amplitude $f(y, z)$ by the expression (8.17). For the last term of (8.17), the amplitude transmitted in the direction $-\theta$ is given to within a constant factor by:

$$f(y, z)f^*(y, z) = |f(y, z)|^2$$

In the direction $-\theta$, the last term of (8.17) gives rise to a plane wave since $|f(y, z)|^2$ is real (no variation of phase). If this plane wave were uniform, it would give rise at the focal point of the objective O_2 to a *luminous point*. In fact this is not so, and the image of the character 木 is replaced not by a single luminous point, which would be the ideal case, but by a luminous point surrounded by a weak and irregular halo. If the holographic filter is not well matched to the character 木, it causes the appearance of a function $g^*(y, z)$ which is different from $f^*(y, z)$. The product $f(y, z)g^*(y, z)$ is not real, and the wavefront diffracted in the direction $-\theta$ is not plane. In the focal plane of the objective O_2, one obtains as response a spot which spreads out much more and becomes still less luminous. One has a spot which becomes less and less visible as the correlation between the character 木 and the character whose Fourier hologram was recorded becomes less and less. This occurs also when the character (the object) does not have a suitable orientation, or when its dimensions are not compatible with the holographic filter which has been constructed.

It is important to note that if in the preceding experiment one translates the character 木 (the object), its transform $f(y, z)$ is multiplied by a factor of the form $e^{jK(u\eta_1 + v\zeta_1)}$, where η_1 and ζ_1 are the coordinates of the new origin with respect to the old one. The last term of (8.17) may be written:

$$ae^{-jK\theta z}e^{jK(u\eta_1 + v\zeta_1)}ff^*$$

and consequently, the wave is still a plane wave but its orientation has changed; it is no longer in the direction $-\theta$. The luminous point changes its position, and it indicates the new position of the character . The luminous point is found in the position occupied by the character 木 if one had been dealing with the first case (the character "object" replaced by the luminous point A_0).

These considerations permit us immediately to solve the second case, when the object A contains the character 木 plus other different signals. The characters different from 木 give practically no "response." Only the characters 木 call attention to their presence by a luminous point which also indicates their positions (Fig. 10.7).

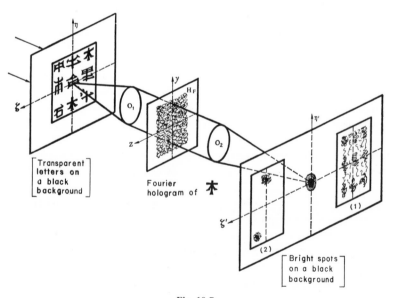

Fig. 10.7

As to the other image situated in the direction θ, it corresponds to an amplitude $f^2(y, z)$ and does not give rise to a signal which is interesting in the present problem.

For each character ⍟ of the object, the hologram transmits amplitudes $f(y, z) f(y, z)$ and $f(y, z) f^*(y, z)$. The amplitude $f(y, z) f(y, z)$ will be found in image (1) and the amplitude $f(y, z) f^*(y, z)$ in the image (2). In order to pass from the hologram to the image plane η', ζ', it is necessary to take the Fourier transform of $f(y, z) f(y, z)$ and of $f(y, z) f^*(y, z)$. Since the magnification between the planes (η, ζ) and (η', ζ') is equal to 1 (Figs. 10.5 and 10.7), one may take the same coordinates for the two planes. The object in the object plane, or its image in the image plane, is represented by the function $F(\eta, \zeta)$ which is the inverse transform of $f(y, z)$. The luminous points which detect in image (2) the presence of the character ⍟ correspond to the transform of the product $f(y, z) f^*(y, z)$, that is to say, to $F(\eta, \zeta) \otimes F^*(-\eta, -\zeta)$ which is the autocorrelation of the function $F(\eta, \zeta)$ (Appendix A, Section A.10). Consequently, the character ⍟ is replaced in the image (2) by the autocorrelation of this character. One observes a luminous point which corresponds to a maximal concentration of light. The correlation $F(\eta, \zeta) \otimes G^*(-\eta, -\zeta)$ of $F(\eta, \zeta)$ with the conjugated function (complex conjugate function) of a different function $G(\eta, \zeta)$ gives in fact a spot which is always more spread out and less luminous.

In the case where the filter is well matched to the object, one then observes the autocorrelation $F(\eta, \zeta) \otimes F^*(-\eta, -\zeta)$ of the function F from which is derived the name of the *method of form recognition by autocorrelation*. In the image (1) one observes $F(\eta, \zeta) \otimes F(\eta, \zeta)$, that is to say, the autoconvolution of the function F. It is interesting to note that if the function F is real and even, there is no difference between the autocorrelation and the autoconvolution, that is to say between the images (1) and (2).

We have taken the example of the detection of a character, but evidently one may detect an ensemble of characters, a phrase or a word by utilizing the Fourier hologram of the ensemble of characters, the phrase or the word.

10.4   Autocorrelation Function in a Simple Case

Take as the signal object the letter O (Fig. 10.8). One may represent this object by a transparent ring on a black background. In Fig. 10.7 it is placed in the plane η, ζ and its Fourier hologram is in the plane y, z. One has a real even object; thus there is no difference between the auto-correlation function and the autoconvolution function. We shall find the same response in the two images (1) and (2). This response is easy to calculate in this case; the autocorrelation function is obtained by calculating the area common to two rings identical to the object as a function of the distance d separating their centers (Fig. 10.9). Figure

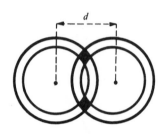

Fig. 10.8   Fig. 10.9

10.10 shows the autocorrelation function, the distance d being expressed as a function of the external diameter D of the letter. We have chosen a ring having a width equal to one-tenth of the diameter D. It is important to note that the ordinates represent amplitudes; thus, in respect of intensities, the decrease will be still more rapid. The response evidently gives rise to a luminous spot standing out against a low-intensity halo. We have a spot whose dimensions are of the order of magnitude of the signal itself (here the letter O). In general, one has a large number of signals which are consequently all very small, compared to the surface area on which they are written. The response corresponding to a given signal will be itself a very small luminous spot, in which one will perceive primarily the central luminous point (Fig. 10.10).

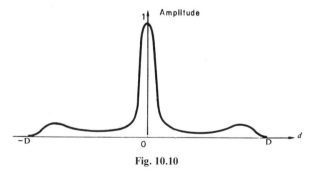

Fig. 10.10

10.5 Detection of Differences between Two Similar Forms

This is the inverse problem to the one considered previously. One method consists of "subtracting" the two photographs I_1 and I_2 whose differences one wish to make evident. If I_1 and I_2 are two positives, one forms a negative copy I'_2 of I_2 and one superimposes I_1 and I'_2 which is equivalent to $I_1 - I_2$.

One may also code the image by means of a grating of step Z_0 placed in contact with the photographic plate (Fig. 10.11). In the first exposure, one records the image $A_1(y, z)$ multiplied by the transmission factor of the grating (Fig. 10.12) which is given by the expression:

$$1 + \frac{4}{\pi}\left[\sin\frac{2\pi z}{Z_0} + \frac{1}{3}\sin 3\frac{2\pi z}{Z_0} + \cdots\right] = 1 + R \qquad (10.1)$$

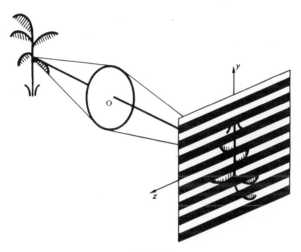

Fig. 10.11

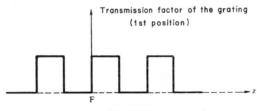

Fig. 10.12

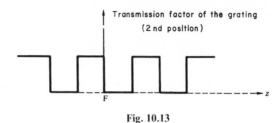

Fig. 10.13

Displace the grating parallel to itself in such a way that the black lines are interchanged with the white lines (Fig. 10.13). In this second position, the transmission factor of the grating is represented by the expression:

$$1 - \frac{4}{\pi}\left[\sin\frac{2\pi z}{Z_0} + \frac{1}{3}\sin 3\frac{2\pi z}{Z_0} + \cdots\right] = 1 - R \qquad (10.2)$$

Record a second image of the object. If this latter has shifted or become deformed, the image is $A_2(y, z) \neq A_1(y, z)$. One has thus recorded the image $A_2(y, z)$ multiplied by the transmission factor of the grating in the complementary position. The photographic plate records two images whose elements are intertwined one with the other. If the object is the same in the two exposures, $A_1 = A_2$ and one has a normal photograph of the object. Simplifying the writing of the formulas, the variations of intensity on the photographic plate may be represented by the expression:

$$A_1(1 + R) + A_2(1 - R) = A_1 + A_2 + (A_1 - A_2)R \qquad (10.3)$$

The difference $A_1 - A_2$ between the two images is modulated at the frequency of the grating, that is to say at a high spatial frequency. If the frequencies of the object are smaller than the frequency of the grating, it is easy to separate $A_1 + A_2$ and $(A_1 - A_2)R$ by filtration. One utilizes the setup of Fig. 10.1, the preceding photograph being placed at A in this figure. A small opaque screen at F intercepts the low frequencies corresponding to the spectrum of $A_1 + A_2$. It passes the high frequencies corresponding to the spectrum of $(A_1 - A_2)R$. In the final image A' (Fig. 10.1) one recovers only the term $(A_1 - A_2)R$ which shows those regions where the two images differ. These regions appear bright against a black background.

Suppose the object is a small, very elongated rectangle, black on a white background (Fig. 10.14) and oriented perpendicularly to the lines of the grating. Between the two exposures the object is displaced, for example, in the direction of the lines of the grating. When one observes the negative in the setup of Fig. 10.1 one sees two images of the object in white on a black background (Fig. 10.15). These images are modulated by the lines of the grating and are complementary: if one superimposes them, the whole field becomes black.

Fig. 10.14

Fig. 10.15

One may study by this method the evolution of atmospheric perturbations photographed by artificial satellites. As another example, one may cite the study of automobile traffic on major highways, using aerial photographs. If one knows the time interval between the two exposures, it is possible to determine the speed distribution of the vehicles. A great number of other applications are possible.

10.6   Information Storage by Means of Holography

One wishes to record N images on a single hologram and to be able later to extract any particular one of these. There are many ways to proceed:

(1)   by making N successive exposures one partially or totally superimposes N images on the hologram,

(2)   by making N successive exposures on N separated regions of the hologram (one region per exposure),

(3)   by making N successive exposures on $N' < N$ separated regions of the hologram (N' regions per exposure).

Figure 10.16 shows an example of the first method in the case where there is partial superposition of the images corresponding to different exposures. The signal to be recorded, for example the character 木, is at A in contact with a diffuser D illuminated by a laser.

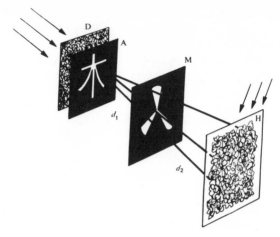

Fig. 10.16

Between the signal to be recorded and the photographic plate H, one places a diaphragm M pierced by an aperture. In the case of Fig. 10.16, the aperture consists of three sectors. The plate H is illuminated also by the reference bundle. After a first exposure, one replaces the character 木 by another signal, for example 土, and one makes a second exposure after turning the sectors so that they occupy a different position which does not reproduce the previous position. One takes a number of exposures in this way, each time changing the signal to be recorded and the position of the sectors. For each exposure it is necessary that the light diffused by D and passing through M shall illuminate all of H reasonably completely. This means that the sectors cannot be too small, and consequently the number of exposures is somewhat limited. Suppose that during the recording one had the arrangement shown in Fig. 10.17. In order to extract a particular chosen signal from the hologram on which the recordings were made,

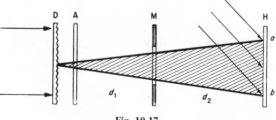

Fig. 10.17

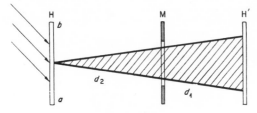

Fig. 10.18

one places this latter at H in Fig. 10.18. The hologram H is turned through 180° with respect to its initial position, and one obtains at H' the (real) signal corresponding to a suitable orientation of the sectors of the diaphragm M. The hologram H reconstructs at M all the sectors utilized during the various recording processes. Consequently, by using a real diaphragm at M to isolate a particular one of these sectors, one reconstructs only the signal which corresponds to it.

It is also noteworthy that if one superposes N signals, the intensity of any one of the reconstructed signals is $1/N^2$ of the intensity which would have been obtained by recording only one image. The signal to noise ratio varies as $1/N^2$.

It is not necessary to go into great detail on the second procedure: N successive exposures of N separated regions of the hologram. This is no different from the recording and observation of an ordinary hologram.

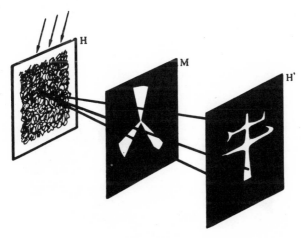

Fig. 10.19

As to the third procedure, it constitutes an intermediate case. The method of recording can be that shown in Fig. 10.16, but now one places the diaphragm in contact with the photographic plate H. One utilizes as many diaphragms as there are exposures to be made; each diaphragm consists of N' small square apertures, all oriented in the same way but distributed at random (Fig. 10.20). If N is the number of

Fig. 10.20

exposures, the number of diaphragms is N/N'. Suppose that the object is uniform and make N exposures; the plate has been illuminated over all its surface. In order to extract a signal, the setup is, for example, that of Fig. 10.18, the diaphragm corresponding to the signal being placed in contact with the hologram (Fig. 10.21). If the number of exposures increases, the resolution diminishes, because for each diaphragm the number and the dimensions of the apertures diminishes.

Other types of apertures may be used: for example, a grating whose orientation is changed from one exposure to the other.

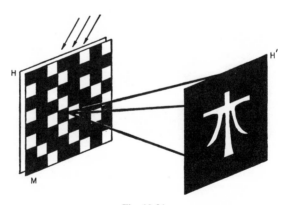

Fig. 10.21

10.7   Attenuation of Noise by Superimposition of Images

Consider an object $O(y, z)$ on which is superposed a noise $B(y, z)$. This may, for example, be a photographic image affected by granularity. Suppose that one has available N copies of the same object on N photographic plates P_1, P_2, P_3, etc. There is no relationship among the noise signals corresponding to the different plates. Record successively on an unexposed plate the images of P_1, P_2, P_3, etc. All the images of the object will be superposed on each other for every exposure. In the final image, the noise is divided by $\sqrt{N}$. Figures 10.22, 10.23, and 10.24 allow us to understand the mechanism of the procedure by considering an object in one dimension. The intensity distribution in any one of the N copies is indicated in Fig. 10.22. Suppose that in the region AB, the noise oscillates around a constant mean value. This mean value represents the intensity $E(z)$ of the image in the region AB. It stays the same regardless of which copy P_l is considered, and it is the deviations from $E(z)$ which will vary. One assumes, in fact, that

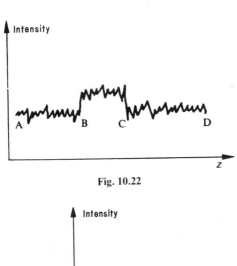

Fig. 10.22

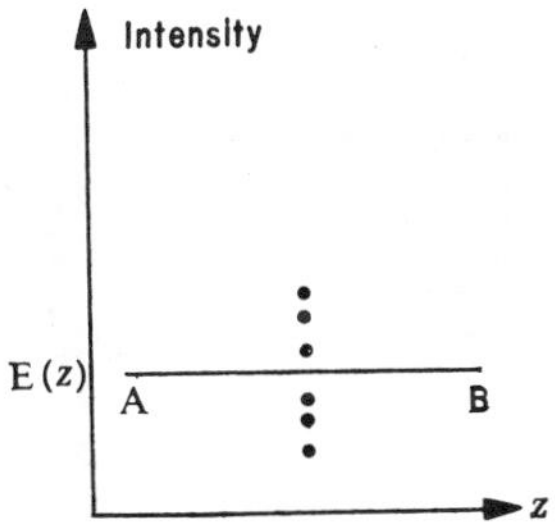

Fig. 10.23

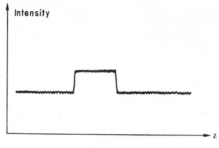

Fig. 10.24

there is no relation whatever between the noise in copies P_1, P_2, P_3, etc. If the number of copies increases, the mean value of the deviations at an arbitrary point of AB diminishes. In Fig. 10.23 we have represented the noise fluctuations corresponding to the abscissa z when one passes from one copy to the next, by means of points on the diagram. One ends up by having as many points above as below the mean value, and this holds true whatever may be the abscissa z of the point considered. The mean value of the displacement with respect to $E(z)$ tends towards zero. The "variance" being the mean of the square of the displacements with respect to the mean, the relationship between the variance and the number N of superpositions shows that the noise is divided by $\sqrt{N}$. One thus "sorts" the signal out of the noise (Fig. 10.24).

10.8   Processing of Images by Computer

In general, calculations are carried out by the computer, starting with digital values of the input data. If the object to be processed consists of continuous variations of intensity, one replaces these continuous variations by discrete values in large enough numbers so that the object can be represented with a sufficient amount of information. The object is digitalized: the coordinates and intensities of the points which make it up are entered into the computer. This latter can now carry out all the operations one wishes: filtering, form recognition, etc. The results of the calculations are received by a digital converter (Fig. 10.25), transformed first into electrical signals XYZ, and finally into a light signal by means of a cathode-ray tube. The signals X and Y determine the position of the luminous spot on the screen of a cathode-ray tube and the signal Z controls the intensity of the spot. In accordance with

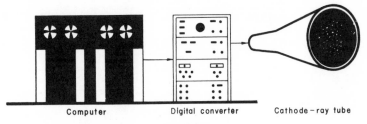

Computer   Digital converter   Cathode-ray tube

Fig. 10.25

the program entered into the computer, one may display on the cathode-ray tube either the spectrum of the image, or the image itself after processing. Note that the cathode-ray tube is not the only method used for displaying images. One also employs mechanical printers which make a trace on a sheet of paper. In all cases, the calculations are based on the fundamental relation (7.2) and (7.3):

transform of the image = (transform of the object)

$$\times \text{ (transform of the image of a point)}$$

$$(10.4)$$

Take as the object a photograph which is degraded by a known defect of focus. The computer carries out the calculation:

$$\frac{\text{transform of the image}}{\text{transform of the image of a point}} = \text{transform of the image} \quad (10.5)$$
$$\text{without defect of focus}$$

In carrying out the preceding calculation (10.5), the computer gives the transform of the image with the defects of focus removed. By an inverse transformation, the computer finally calculates the image without the aberrations introduced by the defects of focus.

Evidently, this processing of the image is possible only if one knows the transform of the image of a point under the conditions of the experiment. The degradation of the image of a point may be due to aberrations of the optical system with which the photograph had been taken, to movement of the object or the plate during the exposure, to nonuniform illumination, to the superposition of a noise such as the granularity or the turbulence of the atmosphere, etc. Most often, one does not know the structure of the image of a point at the moment when the photograph was taken. In analyzing the mechanism which causes the degradation, one may arrive at a concept of the perturbations caused in the ideal image of a point. One utilizes this idea, which is more or less an approximation of reality, in order to carry out the calculations. The operation represented by the relation (10.5) is a filtering operation. It is thus necessary to construct the filter:

$$\frac{1}{\text{transform of the image of a point}}$$

which is possible using calculations on the computer. Things will not be as easy if one tries to construct such a filter optically. One often encounters problems which are difficult, or even impossible, to solve: for example, if the transform of the image of a point has a negative component.

Figure 10.26 represents the application of computer processing of images transmitted by satellites. The space vehicle takes a photograph which is transformed, on board the vehicle, into electrical signals which are transmitted in digital form to the receiving station. The signals are processed by the computer, then transmitted to a converter connected to a system capable of drawing the image; for example, a cathode-ray tube.

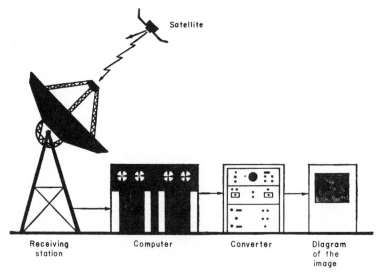

Satellite

Receiving
station

Computer

Converter

Diagram
of the
image

Fig. 10.26

CHAPTER 11

The Optics of Lasers

11.1 Beats Produced by Optical Frequencies

In this chapter, we shall describe certain experiments which can be performed only because of the properties of lasers in regard to temporal coherence and power.

Consider the set up shown in Fig. 11.1. Two lasers L_1 and L_2 simultaneously illuminate a receiver R, a photomultiplier for example,

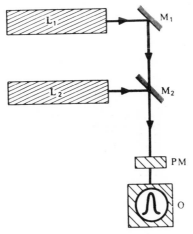

Fig. 11.1

connected to an oscilloscope O. We assume that the coherence time is sufficiently long so that it is possible to make observations during the lifetime of a train of waves. Moreover, we assume that the two lasers emit two wavetrains at almost the same time, in order that the phenomena shall be observable during the time when the two trains are superposed. On arriving at the receiver, the quasi-monochromatic vibrations emitted by L_1 and L_2 may be represented by the expressions (Chapter 5)

$$a_1(t)e^{j2\pi v_1 t} = |a_1(t)|e^{j\varphi_1(t)}e^{j2\pi v_1 t} \tag{11.1}$$

$$a_2(t)e^{j2\pi v_2 t} = |a_2(t)|e^{j\varphi_2(t)}e^{j2\pi v_2 t} \tag{11.2}$$

where $a_1(t)$ and $a_2(t)$ vary slowly compared to v_1 and v_2 and also compared to $v_1 - v_2$. The signal given by the photomultiplier is proportional to:

$$I(t) = (a_1 e^{j2\pi v_1 t} + a_2 e^{j2\pi v_2 t})(a_1^* e^{-j2\pi v_1 t} + a_2^* e^{-j2\pi v_2 t}) \tag{11.3}$$

and, setting $I_1 = a_1 a_1^*$ and $I_2 = a_2 a_2^*$ we have:

$$I(t) = I_1(t) + I_2(t) + 2\sqrt{I_1(t)I_2(t)} \cos[2\pi(v_1 - v_2)t + \varphi_1(t) - \varphi_2(t)]$$

$$\tag{11.4}$$

In accordance with the hypotheses I_1, I_2, φ_1 and φ_2 are practically constant during the lifetime of the train of waves. Consequently, the signal $I(t)$ coming from the photomultiplier is modulated sinusoidally at the frequency $v_1 - v_2$, and one can observe the beats produced between the two optical frequencies v_1 and v_2.

11.2   Interference Phenomena Produced by Two Lasers

The schematic arrangement of the experiment is shown in Fig. 11.2. The two lasers L_1 and L_2 illuminate the two slits T_1 and T_2. After reflection at the two mirrors M_1 and M_2, everything is as if the vibrations originated in T_1' and T_2', which are symmetric to T_1 and T_2 with respect to M_1 and M_2. The lasers emit polarized vibrations, and in order to make them parallel, a polarizer $\mathscr{P}$ is placed in front of the receiver R.

One may return to the expressions (11.1) and (11.2) to represent the vibrations emitted by the two lasers in the plane of the receiver R. We explicitly denote by θ_1 and θ_2 the times taken by the vibrations

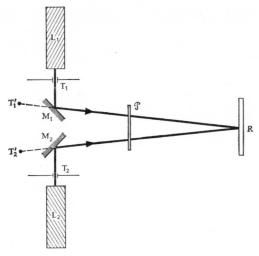

Fig. 11.2

to travel from T'_1 and T'_2 to a general point of the receiver R, and setting $\theta = \theta_1 - \theta_2$, one has for the vibrations (5.5):

$$a_1(t)e^{j2\pi v_1 t}, \qquad a_2(t)e^{j2\pi v_2(t+\theta)} \qquad (11.5)$$

and, once again carrying out the previous calculation, the instantaneous light intensity is:

$$I(t) = I_1(t) + I_2(t)$$
$$+ 2\sqrt{I_1(t)I_2(t)} \ \cos[2\pi(v_2 - v_1)t + \pi v_2\theta + \varphi_1 - \varphi_2] \qquad (11.6)$$

At the instant t the intensity $I(t)$ is not the same at all points of the receiver R. The intensity is a sinusoidal function of θ and one can observe interference fringes. If the frequency $v_1 - v_2$ is sufficiently small so that during the time of an observation the variation of $2\pi(v_1 - v_2)t$ can practically be neglected, the intensity, as a function of θ, will have its maximum when the cosine is equal to $+1$ and its minimum when the cosine is equal to -1, from which the fringe contrast is:

$$\gamma = \frac{2\sqrt{I_1 I_2}}{I_1 + I_2} \qquad (11.7)$$

It is a maximum when $I_1 = I_2$. When two new trains of waves, emitted by L_1 and L_2, interfere, the phases φ_1 and φ_2 have changed. Fringes are visible, but their positions are not the same as before. If the different systems of fringes succeed each other so rapidly that the receiver doesn't have enough time to "see" them separately, the systems of fringes become mixed up, and one no longer sees anything. This is the reason why it is not, in practice, possible to observe interference with two distinct ordinary sources; they emit a great many wavetrains during the time needed to make an observation.

11.3   Nonlinear Optics

In a great many physical experiments, the relationship which exists between two quantities is linear only to a first approximation. Take, for example, the case of a pendulum OM which is displaced through an angle θ from its equilibrium position (Fig. 11.3). The pendulum experiences a force F given by:

$$F = mg \sin \theta = mg\left(\theta - \frac{\theta^3}{3!} + \cdots\right) \tag{11.8}$$

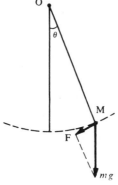

Fig. 11.3

If the angle θ is small, the relation between F and θ is linear; for larger values of θ, however, it is necessary to consider further terms in the development of $\sin \theta$, and the relation between F and θ is no longer linear.

Nonlinear phenomena may also be observed in optics in the relationship between the polarization P of a crystal and the electric field E. One has:

$$P = \chi E(1 + a_1 E + a_2 E^2 + \cdots) \tag{11.9}$$

where x is the polarizability and a_1 is of the order of magnitude of the inverse of the atomic field within the crystal. When a crystal is illuminated by an ordinary source, the field E is never very large, and one writes:

$$P = \chi E \qquad (11.10)$$

In the case of extremely intense fields, it is always necessary to take into account the succeeding terms of the development. One may give an explanation of the nonlinearity by comparing the origin of the electronic polarization of the crystal to the movement of a pendulum. Consider a nucleus surrounded by electrons. If one applies a variable electric field, the nucleus and the electrons are displaced in opposite directions and form a dipole. In fact, considering its mass, the nucleus may be thought of as fixed and the electrons as moving. If the oscillations are of small amplitude, one may assume that there is a proportionality between the force and the displacement, as in the case of the pendulum. This is what happens if the field is weak; but for oscillations of large amplitude produced by intense electric fields, the relation is no longer linear.

Lasers produce intense fields, and are ideal sources for observing nonlinear phenomena. With a ruby laser, one can obtain fields of 10^5 V/cm in the emerging beam. By concentrating the beam on a small surface of diameter 1 μm, the electric field can attain 10^8 V/cm, which is of the order of magnitude of the atomic fields present in the atomic structure of the crystal. In these conditions, the coefficients $a_1, a_2, \ldots$, of equation (11.9) can no longer be neglected; one is dealing with nonlinear optics.

Suppose that only the first two terms of (11.9) are significant in the crystal considered; then one has:

$$P = \chi E + \chi a_1 E^2 \qquad (11.11)$$

Illuminate the crystal with radiation of angular frequency ω. The crystal is subjected to an electric field of the form:

$$E = E_0 \sin \omega t \qquad (11.12)$$

Then (11.11) may be written:

$$P = \chi E_0 \sin \omega t + \tfrac{1}{2} \chi a_1 E_0^2 (1 - \cos 2\omega t) \qquad (11.13)$$

The two terms $\chi E_0 \sin \omega t$ and $\tfrac{1}{2} \chi a_1 E_0^2 \cos 2\omega t$ correspond to two waves of polarization which give rise to two electromagnetic waves of angular frequencies ω and 2ω. On leaving the crystal, we note the presence of the incident radiation of angular frequency ω and a new radiation of angular frequency 2ω, called the second harmonic.

One may perform the experiment by concentrating the beam of a ruby laser on a quartz crystal. If the radiation emitted by the laser has a wavelength of 7000 Å, one finds at the exit point of the quartz plate a very intense beam of the same wavelength, plus a new radiation, not so intense, of wavelength 3500 Å (second harmonic). The proportion of the incident radiation which is transformed into the second harmonic depends on the crystal chosen and the intensity of the incident radiation. The second harmonic wave may be detected by means of a spectrograph.

11.4   Intensity of the Electric Field When It Leaves the Crystal

Consider a crystalline plate with parallel faces and of thickness l (Fig. 11.4). Let $E_0 \sin(\omega t - \varphi_1)$ be the electric field at a distance x from the entry face. One has:

$$\varphi_1 = \frac{2\pi n_1 x}{\lambda} = K_1 x \tag{11.14}$$

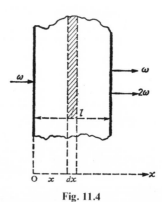

Fig. 11.4

where n_1 is the index of the crystal for the incident radiation of wavelength λ. The electric field due to the wave of polarization $\frac{1}{2}\chi a_1 E_0^2 \cos 2\omega t$, which is the last term on the right-hand side of (11.13), is proportional to:

$$\cos 2(\omega t - \varphi_1) \tag{11.15}$$

A little slice dx of the crystal will produce at the instant t and at the exit surface of the crystal an electric field proportional to:

$$\cos 2[\omega(t - t') - \varphi_1]\, dx \qquad (11.16)$$

where t' is the time required for the radiation $\lambda/2$ (second harmonic) to traverse the path $l - x$ in the crystal. One has:

$$t' = \frac{l - x}{v_2} \qquad (11.17)$$

v_2 being the phase velocity of the second harmonic. Set:

$$v_2 = \frac{2\omega}{K_2} \qquad (11.18)$$

with:

$$K_2 = \frac{2\pi n_2}{\lambda/2} \qquad (11.19)$$

where n_2 is the index of the crystal for the second harmonic. The electric field produced by the second harmonic at the exit from the crystal will be proportional to:

$$\cos[2\omega t - K_2 l + (K_2 - 2K_1)x]\, dx \qquad (11.20)$$

and the field due to the whole thickness of the crystal is:

$$\int_0^l \cos[2\omega t - K_2 l + (K_2 - 2K_1)x]\, dx$$
$$= \frac{2}{K_2 - 2K_1} \sin\left(\frac{K_2}{2} - K_1\right)l \cdot \cos\left[2\omega t - \left(\frac{K_2}{2} + K_1\right)l\right] \qquad (11.21)$$

from which the intensity of the field produced by the second harmonic at the exit of the crystal is

$$I = \left[l\,\frac{\sin[(K_2/2) - K_1]l}{[(K_2/2) - K_1]l}\right]^2 \qquad (11.22)$$

and, replacing K_1 and K_2 by their values as given by (11.14) and (11.19), one has:

$$I = \left\{l\,\frac{\sin[(l\omega/c)(n_2 - n_1)]}{(l\omega/c)(n_2 - n_1)}\right\}^2 \qquad (11.23)$$

where c is the speed of light in a vacuum. This significant result shows the effect of dispersion. If the index n_1 of the crystal for the incident

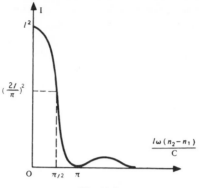

Fig. 11.5

wavelength λ is different from the index n_2 of the crystal for the wavelength $\lambda/2$ (second harmonic), the intensity I of the second harmonic is reduced. The effect of dispersion is therefore always detrimental.

Calculate the intensity of the second harmonic wave for a given value of the dispersion $n_2 - n_1$. Figure 11.5 shows the variation of I as a function of $l\omega(n_2 - n_1)/c$ in accordance with equation (11.23). For $l\omega(n_2 - n_1)/c = \pi/2$ one has moreover:

$$I = \left(\frac{2l}{\pi}\right)^2 \qquad (11.24)$$

This is so if the thickness of the crystal is equal to:

$$l = \frac{\lambda}{4(n_2 - n_1)} \qquad (11.25)$$

l is often called the *coherence length*.

Being given the customary values of the dispersion, l is of the order of twenty or so wavelengths, which is not attainable in practice.

Assuming that it were possible to set $n_1 = n_2$, one then sees that the intensity of the second harmonic wave is proportional to l^2, that is to say, to the square of the thickness of the crystal.

11.5   Method for Increasing the Intensity of the Harmonic. Index Matching

For isotropic crystals, it is not possible to have $n_1 = n_2$ and dispersion is always present. There do exist anisotropic crystals for which, in a given direction of propagation, the ordinary index for one frequency (the incident frequency) is equal to the extraordinary index for the doubled frequency (second harmonic). Figure 11.6 shows the index surface for quartz (see Section 6.4) for the initial frequency (indices n_{o1} and n_{e1}) and for the second harmonic (indices n_{o2} and n_{e2}). The index surfaces referring to ω and 2ω do not intersect; therefore the condition stated earlier cannot be achieved. On the other hand, in the case of KDP (potassium di-hydrogen phosphate) (Fig. 11.7), the surfaces intersect when the direction of propagation is inclined at an angle θ with respect to the optic axis. If the laser beam traverses the crystal in this direction, one has:

$$n_{o1} = n_{e2} \qquad (11.26)$$

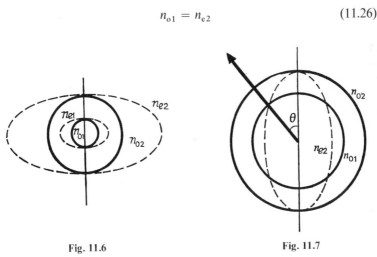

Fig. 11.6   Fig. 11.7

If n_{o1} is the index of the crystal for the incident wave and n_{e2} the index for the second harmonic wave, the condition $n_1 = n_2$ in equation (11.23) is satisfied, and the intensity of the second harmonic wave is a maximum.

11.6 Second Harmonic and Crystal Symmetry

The possibility that a crystal may be used to produce the second harmonic depends on the symmetry properties of the crystal. The coefficient a_1(11.9) vanishes for crystals having, like Iceland spar, a center of inversion (Fig. 11.8). When there is a center of inversion, the polarization changes sign with E and cannot involve even powers of E.

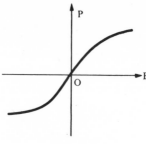

Fig. 11.8

It is the absence of a center of inversion (Fig. 11.9) that makes possible the production of the second harmonic. It is nevertheless possible to make use of a crystal having a center of inversion if one subjects it to a constant intense electric field. This field produces a displacement of the origin from O to O' (Fig. 11.10), and the development of P as a function of E then contains even powers of E. The experiment has been carried out by applying a strong constant electric field to a crystal in a direction perpendicular to the laser beam (Fig. 11.11). The intensity of the second harmonic thus generated is proportional to the square of the field E.

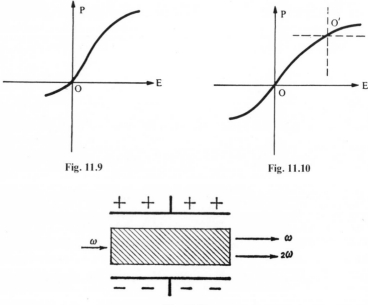

Fig. 11.9   Fig. 11.10

Fig. 11.11

11.7   Basic Concepts of Optical Communication Using Fibers

Beginning in 1960, the laser became the starting point of a great deal of work concerned with using light for the transmission of information. Circuits of a novel type (optical circuits) and a new chapter of optics (integrated optics) made their appearance. It is evident that the higher the frequency one deals with, the greater the quantity of information that can be transmitted. Because of their very high frequencies, of the order of 300,000 gigahertz, light waves are capable of transporting enormous amounts of information. Besides, the radio spectrum is becoming more and more cluttered, and this adds to the interest in using the optical region. The problem is to redo for light waves what has already been done for radio waves. It is necessary to transmit and to process the light signal. Atmospheric absorption will not permit the signal to be transmitted in the open air. Light transmission will have to take place in a medium having favorable properties, and physicists have chosen optical fibers as a transmission medium.

A light ray which penetrates into a fiber undergoes total reflection (Fig. 11.12). It is reflected a great number of times without undergoing any attenuation, other than that caused by the absorption of the material of the fiber, or the diffusion caused by irregularities present within the fiber or on its surface. In general, the fiber is surrounded by a sheath or cladding of index n_2 which is smaller than the index n_1 of the fiber (Fig. 11.13). The sheath or cladding protects the fiber; it allows one to prevent contamination of the surface of the fiber. In cases where numerous fibers are in contact, it eliminates losses from one fiber to another.

The absorption of optical fibers is far from negligible. This is the reason why research on optical communication was not well developed until about 1970, at which time fibers of very small absorption became

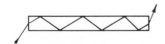

Fig. 11.12   Propagation of light rays in a fiber.

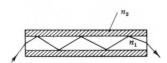

Fig. 11.13   Fiber of index n_1 surrounded by a sheath of index $n_2 (n_2 < n_1)$.

available. The absorption of a fiber, in decibels, is specified by the following formula:

$$A = 10 \log_{10} \frac{\text{incident intensity}}{\text{transmitted intensity}} \qquad (11.27)$$

This is the *optical density* multiplied by 10. An absorption $A = 3$ dB per kilometer length of the fiber, which corresponds to a loss of 50%, may be considered quite small. It is possible nowadays to manufacture fibers having absorption of this order of magnitude, and as a result optical communications has entered into the domain of practical application. Optical fibers take up little space and are light in weight, and they have the advantage of being insensitive to external electromagnetic fields. Their information-carrying capacity is considerable and reaches several hundreds of millions of impulses per second and per fiber.

The wavemodes which propagate in a fiber, which are finite in number, correspond to the zig-zag paths of the rays in Figs. 11.12 and 11.13. These modes are the solution of the propagation equation deduced from Maxwell's equations. Assume that the variation of the index of the fiber has the profile indicated in Fig. 11.14. This is typical

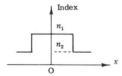

Fig. 11.14   Index profile of the fiber of Figure 11.11.

of the simplest type of fiber. The number N of modes transmitted by such a fiber is given by the formula:

$$N \simeq 2\pi^2 \frac{a^2}{\lambda^2} (n_1^2 - n_2^2) \qquad (11.28)$$

where a is the radius of the fiber, n_1 is the index of the fiber or core, and n_2 is the index of the sheath. The number of modes may be modified by changing the law of variation of the index as a function of the distance from the center of the fiber, that is to say, by modifying the profile. It may be shown that fibers which have a gradual variation in index transmit fewer modes than do fibers having a rectangular profile (11.28). It is very important to know the number of modes, since this determines the information-carrying capacity. When a bundle of light rays penetrates into the fiber, it is decomposed into a number of modes which propagate at different speeds; it is necessary to be prepared for the possibility that the recombination of these modes at the exit of the fiber may not reproduce the signal which entered the fiber. The number of modes is related to the quality of the propagation.

If for a given mode the cut-off frequency is v, the system is incapable of transmitting frequencies greater than v; therefore it will not be possible to transmit two signals separated by a time interval less than $1/v$. From this, one can see the enormous transmission possibilities presented by optical frequencies.

11.8   Waveguides and Optical Components

Fibers form the medium for the transmission of optical information. But the light signal must be processed by optical components which are capable of carrying out the operations of modulation, switching, filtering, etc. These optical components are formed in active or passive thin films, deposited on a substrate in which are located "waveguides." These waveguides are of very small size, as may be seen in the case of the rectangular waveguide shown in Fig. 11.15. In order to achieve

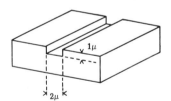

Fig. 11.15   Optical waveguide.

sufficient precision one makes the waveguides using masking techniques and electron beams. Figure 11.16 shows the principle of this method. On the substrate one deposits a thin film of the material which is to be used for making the waveguide, and on top of this a thin film of thermoplastic. The thermoplastic is irradiated in the region where one wishes to construct the waveguide, and is developed like a photographic plate, but in a special solution. If the thermoplastic is "positive," the irradiated region is dissolved during development. Then one has a trough or cavity in this region, as may be seen in Fig. 11.16a. The whole

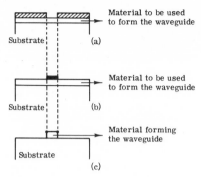

Fig. 11.16   Construction of an optical waveguide.

assembly is metallized by vacuum evaporation, and the remaining thermoplastic is dissolved. There then remains a thin metallic film solely over the region of the material which will become the waveguide (Fig. 11.16b). An ion beam irradiates the assembly, and finally we have a waveguide which stands out in relief (Fig. 11.16c). It is also possible to obtain waveguides in the form of a trough or cavity, and even in the form of a guide involving differences in index of refraction (Fig. 11.17).

Fig. 11.17   Optical waveguide utilizing difference in index.

Figure 11.18 shows an example of an optical component which acts as a switch. It is possible, at a junction between one waveguide and another, to change the transmission of light energy by using the electro-optic effect to modify the index of the material between the two guides. The index of the material employed, for example lithium tantalate, $LiTaO_3$, changes when one changes the voltage between the two

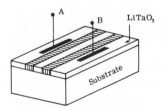

Fig. 11.18   Optical switching.

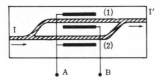

Fig. 11.19   Optical modulator.

electrodes A and B. The component shown in Fig. 11.19 allows one
to construct a light beam modulator, using an optical circuit analogous
to an interferometer. By varying the voltage between the electrodes,
say by means of a microphone, one can change the phase difference
between the two beams which follow the paths (1) and (2); the beam
I' which leaves this structure is modulated by the microphone.

Figure 11.20 shows the principle of a frequency filter. It consists of
two rectangular guides separated by a circular guide. If the circum-
ference of the circular guide is made equal to an integer multiple of the
wavelength, corresponding, for example, to the frequency f_2, the
device behaves like an oscillating circuit for that frequency. If several
frequencies are propagating in guide number 1, only the resonant
frequency f_2 will be transmitted through the circular guide and then
through rectangular guide number 3. One can thus construct frequency
filters by varying the frequency of the oscillating circuit, that is to say,
the dimensions of the circular guide.

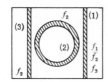

Fig. 11.20   Optical frequency filter.

Finally, we note that it is possible to construct a genuine two-
dimensional optics using thin films. Local variations of index of
refraction behave like variations in thickness, which makes it possible
to construct the equivalent of lenses, prisms, etc.

11.9   Coupling Devices in Integrated Optics

In the preceding sections, we have briefly discussed the properties
of optical components in which the light is processed in various ways.

Fig. 11.21   Optical prism coupler.

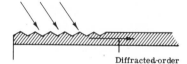

Diffracted-order

Fig. 11.22   Optical grating coupler.

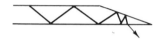

Fig. 11.23   Coupler using a wedge-shaped film.

It is now necessary to consider how to introduce the light energy into the thin films constituting the optical components. Three examples are given in Figs. 11.21, 11.22, and 11.23. Figure 11.21 shows the principle of coupling using a prism. The evanescent wave is "trapped" in the guide: this phenomenon is called the "optical tunnel effect" by analogy with quantum mechanics. In fact, one can show that the field distribution in this case is the same as one finds in the quantum-mechanical problem of a potential step.

Figure 11.22 shows the principle of a grating coupler. The grating is a phase grating designed so that as much as possible of the energy will be concentrated in the order which will propagate in the guide.

In the case of coupling by a wedge-shaped film (Fig. 11.23), the thin film is terminated in a wedge. Because of this, the angle of reflection at the inclined face diminishes, and when the critical angle on the substrate is reached, the wave will be able to refract into it. The wave is introduced into the guide by the inverse optical path.

11.10   Sources and Detectors in Integrated Optics

It is possible to use an ordinary laser and to couple it to a guide. It is preferable, however, to use optical circuits together with a suitable source. One utilizes either diode lasers (coherent sources) or electro-

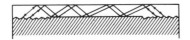

Fig. 11.24   Resonant cavity employing two gratings R_1 and R_2.

luminescent diodes (LED, incoherent sources). The laser effect is obtained in a thin film by means of phase gratings. The principle is shown in Fig. 11.24. The gratings R_1 and R_2, which have suitable profiles, are on the substrate, and one deposits on the assembly a thin film of a material capable of amplification. For other than normal incidence, the gratings R_1 and R_2 behave like mirrors having a large reflection coefficient.

As regards detection, the technology of photodetectors has evolved toward miniaturization, and photodetectors are integrated into optical circuits. They generally consist of germanium or silicon photodiodes. The laser beam first passes into a waveguide. It then may be refracted by a flat prism (local variations of the index of refraction) and then will pass into a modulator controlled for example by a microphone. The speech-modulated beam is transmitted to the fiber by a coupler. At the other end of the fiber, the beam is coupled to a waveguide, which transmits the signal to a detector (Fig. 11.25).

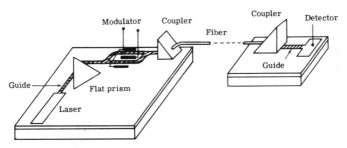

Fig. 11.25   Sketch of the principle of a device for optical communication.

APPENDIX A

Review of Some Elementary Concepts Regarding the Fourier Transformation

A.1 Definition

Let ζ be a real variable which ranges between $-\infty$ and $+\infty$. Consider a function $F(\zeta)$ which may have real or complex values. By definition, the Fourier transform of the function $F(\zeta)$ is the function of the real variable v defined by the expression:

$$f(v) = \int_{-\infty}^{\infty} F(\zeta)e^{j2\pi v\zeta}\, d\zeta, \qquad j = \sqrt{-1} \tag{A.1}$$

One also says that $f(v)$ is the spectrum of $F(\zeta)$. The reciprocity of the Fourier transformation permits one to write:

$$F(\zeta) = \int_{-\infty}^{\infty} f(v)e^{-j2\pi v\zeta}\, dv \tag{A.2}$$

The Fourier transformation is linear: the transform of a sum of functions is the sum of the transforms. More generally, the transform of a linear combination of functions is the same linear combination of the transforms of these functions.

A.2   Translation

If one translates the function $F(\zeta)$ by a displacement ζ_0 its transform is simply multiplied by $e^{j2\pi v\zeta_0}$:

$$\int_{-\infty}^{\infty} F(\zeta - \zeta_0)e^{j2\pi v\zeta}\,d\zeta = e^{j2\pi v\zeta_0}f(v) \tag{A.3}$$

This result is easily proved by setting $\zeta' = \zeta - \zeta_0$. In the same way:

$$\int_{-\infty}^{\infty} f(v - v_0)e^{-j2\pi v\zeta}\,dv = e^{-j2\pi v_0\zeta}F(\zeta) \tag{A.4}$$

A.3   Dilatation

Let a be a real positive constant ($a > 0$); then one has:

$$\int_{-\infty}^{\infty} F(a\zeta)e^{j2\pi v\zeta}\,d\zeta = \frac{1}{a}f\left(\frac{v}{a}\right) \tag{A.5}$$

If a is a real negative constant ($a < 0$):

$$\int_{-\infty}^{\infty} F(a\zeta)e^{j2\pi v\zeta}\,d\zeta = -\frac{1}{a}f\left(\frac{v}{a}\right) \tag{A.6}$$

If $a = -1$, one has:

$$\int_{-\infty}^{\infty} F(-\zeta)e^{j2\pi v\zeta}\,d\zeta = f(-v) \tag{A.7}$$

A.4   Case of Two Variables

If the function $f(\mu, v)$ is the Fourier transform of $F(\eta, \zeta)$, one has:

$$f(\mu, v) = \iint_{-\infty}^{\infty} F(\eta, \zeta)e^{j2\pi(\mu\eta + v\zeta)}\,d\eta\,d\zeta \tag{A.8}$$

with:

$$F(\eta, \zeta) = \iint_{-\infty}^{\infty} f(\mu, v)e^{-j2\pi(\mu\eta + v\zeta)}\,d\mu\,dv \tag{A.9}$$

A.5 Some Common Fourier Transforms

"Rectangle" function in one dimension ("slit" function). This is equal to 1 or 0, depending on whether $|\zeta|$ is less than or greater than $\zeta'/2$ (Fig. A.1). One may write:

$$F(\zeta) = \text{Rect}\left(\frac{\zeta}{\zeta'}\right) = 1, \qquad |\zeta| < \frac{\zeta'}{2}$$

$$F(\zeta) = 0, \qquad\qquad |\zeta| > \frac{\zeta'}{2} \qquad \text{(A.10)}$$

Its transform is, in normalized form

$$f(v) = \frac{\sin \pi v \zeta'}{\pi v \zeta'} = \text{sinc}(\pi v \zeta') \qquad \text{(A.11)}$$

"Rectangle" function in two dimensions. This is defined by:

$$F(\eta, \zeta) = \text{Rect}\left(\frac{\zeta}{\zeta'}\right) \text{Rect}\left(\frac{\eta}{\eta'}\right) = 1, \qquad |\zeta| < \frac{\zeta'}{2}, \quad |\eta| < \frac{\eta'}{2} \qquad \text{(A.12)}$$

$$F(\eta, \zeta) = 0 \text{ everywhere else}$$

Its transform is, after normalization (Fig. A.2):

$$f(\mu, v) = \frac{\sin \pi v \zeta'}{\pi v \zeta'} \cdot \frac{\sin \pi \mu \eta'}{\pi \mu \eta'} = \text{sinc}(\pi v \zeta') \, \text{sinc}(\pi \mu \eta') \qquad \text{(A.13)}$$

"Circle" function. It is defined by (Fig. A.3):

$$F(\eta, \zeta) = 1, \qquad \eta^2 + \zeta^2 < a_0^2$$

$$F(\eta, \zeta) = 0, \qquad \eta^2 + \zeta^2 > a_0^2 \qquad \text{(A.14)}$$

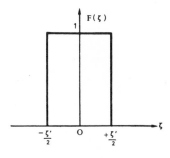

Fig. A.1

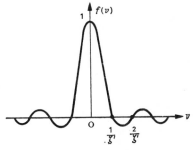

Fig. A.2

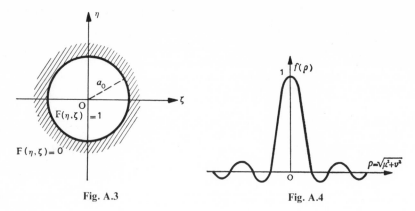

Fig. A.3   Fig. A.4

The transform of the "circle" function is (Fig. A.4):

$$f(\rho) = \frac{2J_1(Z)}{Z}, \qquad \rho = \sqrt{\mu^2 + v^2}, \qquad Z = 2\pi\rho a_0 \qquad (A.15)$$

Limited function $\cos 2\pi v_0 \zeta$. It is defined (Fig. A.5) by:

$$F(\zeta) = \cos 2\pi v_0 \zeta, \qquad -\frac{\zeta_0}{2} < \zeta < +\frac{\zeta_0}{2} \qquad (A.16)$$

and its transform (Fig. A.6) is:

$$f(v) = \text{sinc}[\pi(v + v_0)\zeta_0] + \text{sinc}[\pi(v - v_0)\zeta_0] \qquad (A.17)$$

Limited function $\sin 2\pi v_0 \zeta$. It is defined (Fig. A.7) by

$$F(\zeta) = \sin 2\pi v_0 \zeta, \qquad -\frac{\zeta_0}{2} < \zeta < \frac{\zeta_0}{2} \qquad (A.18)$$

and its transform (Fig. A.8) is, to within a factor $\sqrt{-1}$:

$$f(v) = \text{sinc}[\pi(v - v_0)\zeta_0] - \text{sinc}[\pi(v + v_0)\zeta_0] \qquad (A.19)$$

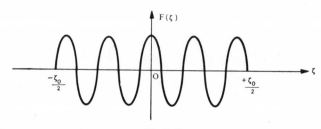

Fig. A.5

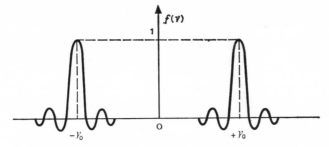

Fig. A.6

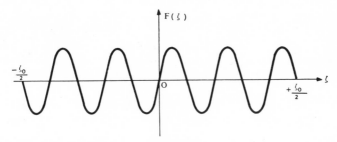

Fig. A.7

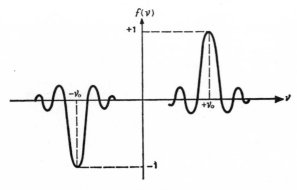

Fig. A.8

Limited function $\cos^2(\pi v_0 \zeta)$. It is defined (Fig. A.9) by

$$F(\zeta) = \cos^2 \pi v_0 \zeta, \qquad -\frac{\zeta_0}{2} < \zeta < \frac{\zeta_0}{2} \qquad \text{(A.20)}$$

and its transform (Fig. A.10) is:

$$f(v) = \text{sinc}(\pi v \zeta_0) + \tfrac{1}{2}\,\text{sinc}[\pi(v + v_0)\zeta_0] + \tfrac{1}{2}\,\text{sinc}[\pi(v - v_0)\zeta_0] \qquad \text{(A.21)}$$

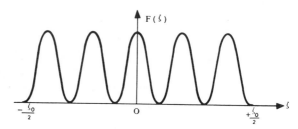

Fig. A.9

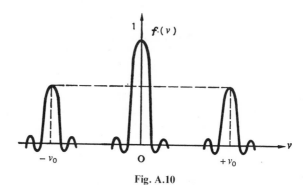

Fig. A.10

A.6   Useful Relations

If $F(\zeta)$ is real:

$$F^*(\zeta) = F(\zeta), \qquad f^*(-v) = f(v), \qquad f^*(v) = f(-v) \quad \text{(A.22)}$$

If $F(\zeta)$ is real and even:

$$F(\zeta) = F^*(\zeta) = F^*(-\zeta), \qquad f(v) = f^*(-v) = f^*(v) \quad \text{(A.23)}$$

If $F(\zeta)$ is real and odd:

$$F(\zeta) = -F(-\zeta) = F^*(\zeta), \qquad f(v) = -f(-v) = f^*(-v) \quad \text{(A.24)}$$

A.7   Convolution

Consider two functions $F_1(\zeta)$ and $F_2(\zeta)$. By definition, the convolution of the two functions $F_1(\zeta)$ and $F_2(\zeta)$ is given by the expression:

$$\int_{-\infty}^{\infty} F_1(\zeta')F_2(\zeta - \zeta')\, d\zeta' \qquad (A.25)$$

and one writes symbolically:

$$F_1(\zeta) \otimes F_2(\zeta) \qquad (A.26)$$

Figures A.11–A.13 summarize the principle of forming the convolution of the two functions $F_1(\zeta)$ and $F_2(\zeta)$. To determine the convolution corresponding to ζ, one translates one of the functions $F_1(\zeta')$ with respect to the other $F_2(\zeta - \zeta')$ and then one forms the product $F_1(\zeta')F_2(\zeta - \zeta')$. The convolution corresponding to ζ is given by the shaded area in Fig. A.13. The calculation is carried out in this way for various values of ζ (see the example given in Section A.12).

One may change the order of factors in the calculation and write:

$$F_1(\zeta) \otimes F_2(\zeta) = F_2(\zeta) \otimes F_1(\zeta) \qquad (A.27)$$

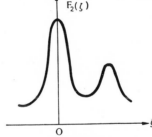

Fig. A.11   **Fig. A.12**

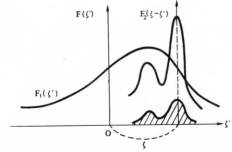

Fig. A.13

A.8   Transform of a Convolution

Let $f_1(v)$ and $f_2(v)$ be the transforms of $F_1(\zeta)$ and $F_2(\zeta)$. The following relations are valid:

$$\text{F.T.}[F_1 \otimes F_2] = f_1(v)f_2(v)$$
$$\text{F.T.}[F_1 F_2] = f_1(v) \otimes f_2(v) \tag{A.28}$$

Here F.T. signifies the Fourier transform of the expression within the parentheses. The first of the two relations (A.28) is often given the name of Parseval's theorem.

A.9   Cross Correlation

The cross correlation of two functions $F_1(\zeta)$ and $F_2(\zeta)$ is defined by the expression:

$$\int_{-\infty}^{\infty} F_1(\zeta')F_2^*(\zeta' - \zeta)\, d\zeta' = F_1(\zeta) \otimes F_2^*(-\zeta) \tag{A.29}$$

ζ' being a variable of integration. One has:

$$\text{F.T.}[F_1(\zeta) \otimes F_2^*(-\zeta)] = f_1(v)f_2^*(v) \tag{A.30}$$

or, in a slightly different form:

$$\text{F.T.}[F_1(\zeta) \otimes F_2^*(\zeta)] = f_1(v)f_2^*(-v) \tag{A.31}$$

A.10   Autocorrelation

The autocorrelation function of the function $F(\zeta)$ is defined by the expression:

$$\int_{-\infty}^{+\infty} F(\zeta')F^*(\zeta' - \zeta)\, d\zeta \tag{A.32}$$

which is written symbolically:

$$F(\zeta) \otimes F^*(-\zeta) \tag{A.33}$$

The transform of the autocorrelation function is equal to the square of the modulus of the transform $f(v)$ of $F(\zeta)$:

$$\text{F.T.}[F(\zeta) \otimes F^*(-\zeta)] = f(v)f^*(v) = |f(v)|^2 \tag{A.34}$$

In the case of a function $F(\zeta)$ which is real and even, the autocorrelation function may be written simply:

$$F(\zeta) \otimes F(\zeta) \qquad (A.35)$$

and one has:

$$\text{F.T.}[F(\zeta) \otimes F(\zeta)] = |f(v)|^2 \qquad (A.36)$$

Consider the example of a slit function (Fig. A.14) whose Fourier transform is represented in Fig. A.15. The autocorrelation function is determined immediately by calculating the variations of the shaded area in Fig. A.16 as a function of ζ. We have here designated the width of the slit as ζ_0' in order not to cause any confusion with the variable of integration ζ'. Figure A.17 represents the autocorrelation function of $F(\zeta)$ whose transform $|f(v)|^2$ is given in Fig. A.18.

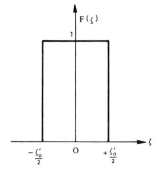

Fig. A.14

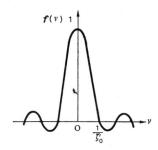

Fig. A.15

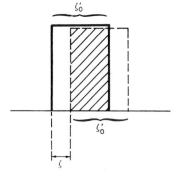

Fig. A.16

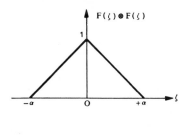

Fig. A.17

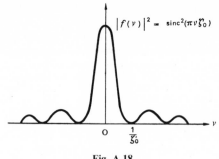

Fig. A.18

A.11   Dirac Distribution

Consider the slit function (Fig. A.19). If $\zeta' \to 0$ and $F(\zeta) \to \infty$ in such a way that the area under the curve remains equal to unity, we call this in the limit the Dirac distribution or delta function $\delta(\zeta)$. One represents $\delta(\zeta)$ by a vertical line of height normalized to unity (Fig. A.20). Its Fourier transform is $f(v) = 1$ (Fig. A.21):

$$\text{F.T.}[\delta(\zeta)] = 1 \qquad (A.37)$$

and if one translates the delta function by an amount ζ_0 one has:

$$\text{F.T.}[\delta(\zeta - \zeta_0)] = e^{j2\pi v\zeta_0} \qquad (A.38)$$

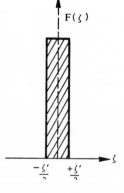

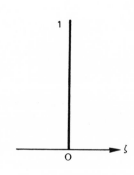

Fig. A.19   **Fig. A.20**

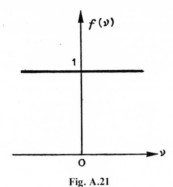

Fig. A.21

In the case of a space of two dimensions, one may define the delta function starting from the circle function:

$$\delta(\eta, \zeta) = \lim_{\varepsilon \to 0} \frac{1}{\pi \varepsilon^2} \text{circle}\left(\sqrt{\frac{\eta^2 + \zeta^2}{\varepsilon^2}}\right) \tag{A.39}$$

and one has:

$$\delta(a\eta, b\zeta) = \frac{1}{|ab|} \delta(\eta, \zeta) \tag{A.40}$$

The delta function may be defined in an analogous way in a space of more than two dimensions.

A.12 Convolution of a Function $F(\zeta)$ with a Delta Function

One has:

$$F(\zeta) \otimes \delta(\zeta) = F(\zeta) \tag{A.41}$$

The delta function is the neutral element, the identity element, of a convolution, just as unity is the neutral element of multiplication.

Subject the function $F(\zeta)$ to a translation ζ_0. One may write:

$$F(\zeta - \zeta_0) = F(\zeta) \otimes \delta(\zeta - \zeta_0) \tag{A.42}$$

The translation may thus be considered as a convolution. Take the transform of both sides of (A.42); according to (A.38) one has:

$$\text{F.T.}[F(\zeta - \zeta_0) = e^{j2\pi v \zeta_0} \text{F.T.}[F(\zeta)] = e^{j2\pi v \zeta_0} f(v) \tag{A.43}$$

a result which has already been obtained in Section A.2.

A.13   Calculation of a Convolution and of a Correlation
in Two Dimensions

One wishes to calculate the convolution of the two functions
$F_1(\eta, \zeta)$ and $F_2(\eta, \zeta)$:

$$F_1(\eta, \ \zeta) \otimes F_2(\eta, \ \zeta) = \iint F_1(\eta', \ \zeta')F(\eta - \eta', \ \zeta - \zeta') \ d\eta' \ d\zeta' \quad \text{(A.44)}$$

where η' and ζ' are two variables of integration.

For simplicity's sake, we take two real functions which are represented in Figs. A.22 and A.23. Turn the curve representing $F_2(\eta', \zeta')$
through 180° in order to get the function $F_2(-\eta', \ -\zeta')$ (Fig. A.24).
Translate $F_2(-\eta', \ -\zeta')$ in such a way that the representative curve of

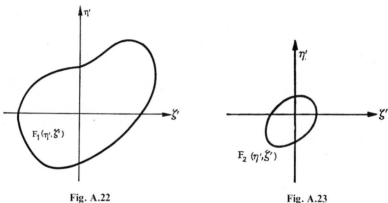

Fig. A.22   Fig. A.23

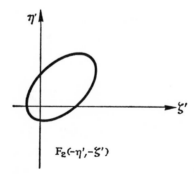

Fig. A.24

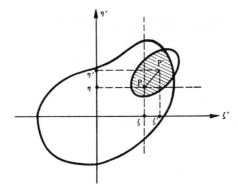

Fig. A.25

this function is referred to axes having an origin P located at the coordinates η and ζ (Fig. A.25). With respect to these new axes, the function F_2 may be written $F_2(\eta - \eta', \zeta - \zeta')$. Therefore, to calculate at the point P the product $F_1(\eta', \zeta')F_2(\eta - \eta', \zeta - \zeta')$ it is necessary to form the product of $F_1(\eta', \zeta')$ at P' with $F_2(\eta - \eta', \zeta - \zeta')$, which represents the value at $P'(\eta', \zeta')$ of the function F_2 referred to axes whose origin is at the point $P(\eta, \zeta)$. The shaded area represents the convolution evaluated at P. Figures A.26–A.28 give as an example the autoconvolution of a function consisting of three delta functions.

Now calculate the following cross-correlation:

$$F_1(\eta, \zeta) \otimes F_2^*(-\eta, -\zeta) = \iint F_1(\eta', \zeta')F_2^*(\eta' - \eta, \zeta' - \zeta)\, d\eta'\, d\zeta'$$

(A.45)

For simplicity one again restricts oneself to the two real functions of Figs. A.22 and A.23. This time it is not necessary to rotate the curve

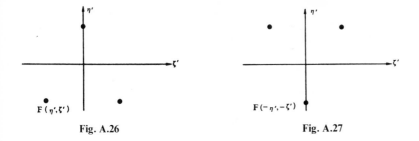

Fig. A.26   **Fig. A.27**

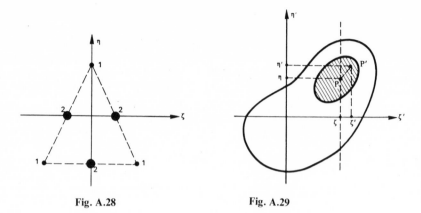

Fig. A.28   Fig. A.29

representing the function F_2. One translates this function in such a way that its representative curve is referred to axes of origin $P(\eta, \zeta)$ (Fig. A.29). Referred to these new axes, the function F_2 is written $F_2(\eta' - \eta, \zeta' - \zeta)$. In order to calculate at the point P the product $F_1(\eta', \zeta')F_2(\eta' - \eta, \zeta' - \zeta)$, it is necessary to form the product of $F_1(\eta', \zeta')$ at $P'(\eta', \zeta')$ with $F_2(\eta' - \eta, \zeta' - \zeta)$ which represents the value at $P'(\eta', \zeta)$ of the function F_2 referred to axes whose origin is at $P(\eta, \zeta)$. Figure A.30 gives the example of the autocorrelation of the function represented in Fig. A.26 (three delta functions).

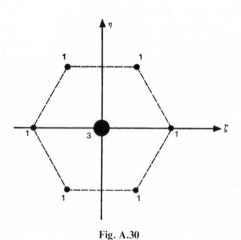

Fig. A.30

One may note that, in the convolution, it is not necessary to change the signs in the function F_2. One cannot write $F_2(\eta' - \eta, \; \zeta' - \zeta)$. The same holds true for the function F_2 of the correlation. On the other hand, the signs of η and ζ in the convolution and the correlation are not important.

A.14   Poisson Distribution or Dirac Comb

The *Dirac comb* (Fig. A.31) of step ζ_0 is defined by:

$$\text{comb}\left(\frac{\zeta}{\zeta_0}\right) = \sum_{n=-\infty}^{+\infty} \delta(\zeta - n\zeta_0) \tag{A.46}$$

Its Fourier transform is:

$$\text{F.T.}\left[\text{comb}\left(\frac{\zeta}{\zeta_0}\right)\right] = \sum_{n=-\infty}^{+\infty} \delta\left(\nu - \frac{n}{\zeta_0}\right) \tag{A.47}$$

The transform of a Dirac comb of step ζ_0 is another comb of step $1/\zeta_0$ (Fig. A.32). One also writes this as $\text{comb}(\nu/\nu_0)$.

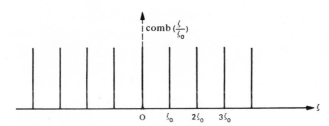

Fig. A.31

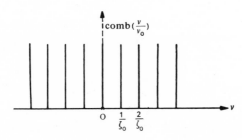

Fig. A.32

A.15   Periodic Function

Let $F(\zeta)$ be an *unbounded* periodic function of period ζ_0 (Fig. A.33). One may consider that the function $F(\zeta)$ is obtained by translating the pulse $\mathscr{F}(\zeta)$ (Fig. A.34) through integer multiples of the distance ζ_0. Since the translation is a convolution, we may write:

$$F(\zeta) = \mathscr{F}(\zeta) \otimes \sum_{n=-\infty}^{+\infty} \delta(\zeta - n\zeta_0) \tag{A.48}$$

$$F(\zeta) = \mathscr{F}(\zeta) \otimes \mathrm{comb}\!\left(\frac{\zeta}{\zeta_0}\right) \tag{A.49}$$

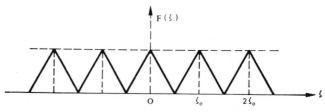

Fig. A.33

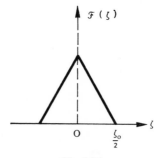

Fig. A.34

One thus immediately obtains the transform of $F(\zeta)$ by taking the transform of the two members of (A.49). One forms the product of the transforms of $\mathscr{F}(\zeta)$ and $\mathrm{comb}(\zeta/\zeta_0)$ from which we arrive at Fig. A.35. Suppose now that the periodic function $F(\zeta)$ were bounded.

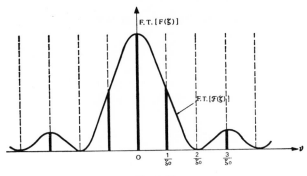

Fig. A.35

We may consider this bounded function $F_L(\zeta)$ as the product of $F(\zeta)$ with a rectangle function of "width" L and, from (A.49):

$$F_L(\zeta) = F(\zeta) \operatorname{Rect}\left(\frac{\zeta}{L}\right) = \left[\mathscr{F}(\zeta) \otimes \operatorname{comb}\left(\frac{\zeta}{\zeta_0}\right)\right] \operatorname{Rect}\left(\frac{\zeta}{L}\right) \quad \text{(A.50)}$$

The Fourier transform of $F_L(\zeta)$ is given by:

$$\text{F.T.}[F_L(\zeta)] = \left\{[\text{F.T. } \mathscr{F}(\zeta)] \operatorname{comb}\left(\frac{v}{v_0}\right)\right\} \otimes \frac{\sin \pi v L}{\pi v L} \quad \text{(A.51)}$$

APPENDIX B

The Fresnel–Kirchhoff Formula
and the Phenomena of Diffraction

B.1 The Helmholtz Equation

At a general point, the light vibration may be represented by the scalar function (1.1) which satisfies the wave equation:

$$\nabla^2 E - \frac{1}{c^2}\frac{\partial^2 E}{\partial t^2} = 0 \qquad (B.1)$$

Set:

$$U = E_m e^{-j\varphi} \qquad (B.1a)$$

U is the complex amplitude of the vibration. Substituting (1.8) and (B.1a) into (B.1) one has, if $K = 2\pi/\lambda$:

$$(\nabla^2 + K^2)U = 0 \qquad (B.2)$$

This is the Helmholtz equation.

B.2 Green's Theorem

The theory of the diffraction of scalar waves is based on Green's theorem. Consider two functions U and G which depend only on the position of the point considered. Consider a closed surface S bounding a volume v. By hypothesis, the two functions U and G have no singularity in the interior of the volume v. At each point of the surface S there is a well-defined normal n, oriented positively toward the exterior of S. Green's theorem may be written:

$$\iiint_v (G\nabla^2 U - U\nabla^2 G)\,dv = \iint_S \left(G\frac{\partial U}{\partial n} - U\frac{\partial G}{\partial n}\right) dS \qquad (B.3)$$

B.3   The Helmholtz–Kirchhoff Integral

The problem which we wish to consider is the following: to calculate the complex amplitude at a point P (Fig. B.1) knowing the amplitude at all the points of a surface S surrounding the point P. Let us imagine that the point P emits spherical waves, and let us represent the amplitude at a distance r from P by the function G:

$$G = \frac{e^{-jKr}}{r} \qquad (B.4)$$

This function has a singularity at P, where $r = 0$. In order to fulfill the conditions of validity of Green's theorem, it is necessary to exclude P from the domain of integration; for this one surrounds P by a small sphere of surface S_ε and radius ε. One applies Green's theorem to the volume v' contained between S and S_ε. The surface of integration is $S' = S + S_\varepsilon$. In the interior of v', the function G represents a progressive spherical wave which satisfies the Helmholtz equation:

$$(\nabla^2 + K^2)G = 0 \qquad (B.5)$$

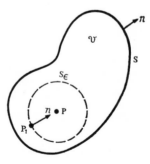

Fig. B.1

Substitute (B.2) and (B.5) in the left-hand side of (B.3); then one has:

$$\iiint_{v'} (G \nabla^2 U - U \nabla^2 G)\, dv = - \iiint_{v'} (GUK^2 - UGK^2)\, dv = 0 \quad \text{(B.6)}$$

from which:

$$\iint_{S'} \left(G \frac{\partial U}{\partial n} - U \frac{\partial G}{\partial n} \right) dS = 0 \quad \text{(B.7)}$$

that is to say:

$$\iint_{S} \left(G \frac{\partial U}{\partial n} - U \frac{\partial G}{\partial n} \right) dS = - \iint_{S_\varepsilon} \left(G \frac{\partial U}{\partial n} - U \frac{\partial G}{\partial n} \right) dS \quad \text{(B.8)}$$

From (B.4) one has:

$$\frac{\partial G}{\partial n} = -\cos(n, v)\left(jK + \frac{1}{r} \right) \frac{e^{-jKr}}{r} \quad \text{(B.9)}$$

$\cos(n, r)$ is the cosine of the angle between the normal n and the vector r joining P to a general point. For a point P_1 situated on S_ε one has $\cos(n, r) = -1$ because, when one applies Green's theorem to the volume contained between S and S_ε, the normal to S_ε is directed toward P; consequently (B.4) and (B.9) must be written:

$$G = \frac{e^{-jK\varepsilon}}{\varepsilon}, \qquad \frac{\partial G}{\partial n} = \left(\frac{1}{\varepsilon} + jK \right) \frac{e^{-jK\varepsilon}}{\varepsilon} \quad \text{(B.10)}$$

Calculate the integral of the right-hand side of (B.8). To the surface element dS there corresponds at P the solid angle $d\Omega = dS/\varepsilon^2$. Then we have:

$$\iint_{S_\varepsilon} \left(G \frac{\partial U}{\partial n} - U \frac{\partial G}{\partial n} \right) dS = \iint_{S_\varepsilon} \left[\frac{e^{-jK\varepsilon}}{\varepsilon} \frac{\partial U}{\partial \eta} - \frac{U e^{-jK\varepsilon}}{\varepsilon} \left(\frac{1}{\varepsilon} + jK \right) \right] \varepsilon^2\, d\Omega$$

$$\text{(B.11)}$$

Allow ε to approach 0; then:

$$\iint_{S_\varepsilon} U\, d\Omega = 4\pi U_P \quad \text{(B.12)}$$

where U_P is the value of U at P. Substituting this in (B.8), it becomes:

$$U_P = \frac{1}{4\pi} \iint_S \left\{ \frac{\partial U}{\partial n} \left(\frac{e^{-jKr}}{r} \right) - U \frac{\partial}{\partial n} \left(\frac{e^{-jKr}}{r} \right) \right\} dS \qquad (B.13)$$

which is the Helmholtz–Kirchhoff integral.

B.4 Diffraction by an Aperture Pierced in a Plane Screen. The Kirchhoff Limiting Conditions

The screen D, containing an arbitrary aperture T (Fig. B.2), is illuminated by a point source L. We wish to calculate the amplitude at a general point P located on the other side of D. Make use of the Helmholtz–Kirchhoff integral. In order to do this, choose a closed surface of integration formed by the following two surfaces:

    (a)   a portion of a spherical surface S_1 of radius R centered at P,
    (b)   a portion of a plane S_2 immediately behind D.

The surface of integration is $S_1 + S_2$ and, from (B.13), one has:

$$U_P = \frac{1}{4\pi} \iint_{S_1+S_2} \left\{ \frac{\partial U}{\partial n} \left(\frac{e^{-jKr}}{r} \right) - U \frac{\partial}{\partial n} \left(\frac{e^{-jKr}}{r} \right) \right\} dS \qquad (B.14)$$

where $r = PM$, M being a general point of the aperture T. In order to calculate the amplitude U_P at P, it is necessary to know the amplitude

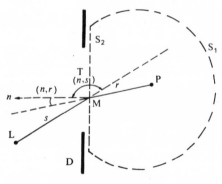

Fig. B.2

U and its derivative $\partial U/\partial n$ at all the points of $S_1 + S_2$. Unfortunately, these quantities are never known exactly. The following hypotheses will permit us to simplify the problem:

(a) U and $\partial U/\partial n$ are taken to be zero on the screen D,

(b)   these two quantities have in the aperture T the same values as in the absence of the screen D.

These simplifications permit us to calculate the integral over the domain S_2, but there remains the integration over S_1. For this, we shall assume that the source L begins to emit a train of waves at the instant t_0. This implies that the source L is not rigorously monochromatic as we had assumed, because a monochromatic vibration is a vibration which exists for all times. At the time t, the vibration emitted by L has traversed the distance $c(t - t_0)$ and if R is chosen sufficiently large, when the vibration reaches the point P, the vibratory movement has not yet reached S_1. In these circumstances, S_1 cannot make any contribution to the vibratory movement at P, and the integral (B.14) relating to the domain of integration S_1 vanishes. One then has:

$$U_P = \frac{1}{4\pi} \iint_T \left\{ \frac{\partial U}{\partial n} \left(\frac{e^{-jKr}}{r} \right) - U \frac{\partial}{\partial n} \left(\frac{e^{-jKr}}{r} \right) \right\} dS \qquad \text{(B.15)}$$

B.5   Fresnel–Kirchhoff Formula

Suppose that the distance r is very much larger than the wavelength, then:

$$\frac{\partial}{\partial n} \left(\frac{e^{-jKr}}{r} \right) \cong - jK \cos(n, r) \frac{e^{-jKr}}{r} \qquad \text{(B.16)}$$

and, substituting into (B.15)

$$U_P = \frac{1}{4\pi} \iint_r \frac{e^{-jKr}}{r} \left[\frac{\partial U}{\partial n} + jKU \cos(n, r) \right] dS \qquad \text{(B.17)}$$

The screen D being illuminated by a point source L (Fig. B.2), the vibration at M is:

$$F(M) = \mathscr{A} \frac{e^{-jKs}}{s} \qquad \text{(B.18)}$$

We may now replace U by $F(M)$ in (B.17) and if $s \gg \lambda$:

$$U_P = \frac{j\mathscr{A}}{\lambda} \iint\limits_{T} \frac{e^{-jK(r+s)}}{rs} \left[\frac{\cos(n, r) - \cos(n, s)}{2} \right] dS \qquad (B.19)$$

which is the Fresnel–Kirchhoff formula. Set:

$$H(M) = \frac{j}{\lambda} \left(\frac{\mathscr{A}e^{-jKs}}{s} \right) \frac{\cos(n, r) - \cos(n, s)}{2} \qquad (B.20)$$

Then one has:

$$U_P = \iint\limits_{T} H(M) \frac{e^{-jKr}}{r} dS \qquad (B.21)$$

The amplitude at P may be considered as due to an infinite number of fictitious secondary sources located in the aperture T. At a general point M, the amplitude of the secondary source is $H(M)$. From (B.20) we see that it is proportional to the amplitude of the incident wave produced by L and in quadrature with it.

If one chooses the experimental circumstances such that (n, r) is small and (n, s) is close to 180°, then:

$$U_P = \frac{j\mathscr{A}}{\lambda} \iint\limits_{T} \frac{e^{-jK(r+s)}}{rs} dS = \frac{j}{\lambda} \iint\limits_{T} F(M) \frac{e^{-jKr}}{r} dS \qquad (B.22)$$

Brief Bibliography

Born, M., and Wolf, E., "Principles of Optics," 5th ed. Pergamon Press, London and New York, 1975.

Ditchburn, R. W., "Light," 3rd ed. Academic Press, New York, 1976.

Françon, M., "Modern Applications of Physical Optics," Wiley, New York, 1963.

Françon, M., "Holography," Academic Press, New York, 1974.

Goodman, J. W., "Introduction to Fourier Optics," McGraw-Hill, New York, 1968.

Mertz, L., "Transformations in Optics," Wiley, New York, 1965.

Index

E

F

G